Elementary!
Puzzles for the Chemically Curious and the Periodically Perplexed

Elementary!: Puzzles for the Chemically Curious and the Periodically Perplexed

By

Paul Board
Email: boardp@gmail.com

Print ISBN: 978-1-83916-945-8

A catalogue record for this book is available from the British Library

The Royal Society of Chemistry is a charity, registered in England and Wales, Number 207890, and a company incorporated in England by Royal Charter (Registered No. RC000524), registered office: Burlington House, Piccadilly, London W1J 0BA, UK, Telephone: +44 (0) 20 7437 8656.

Visit our website at www.rsc.org/books

Printed in the United Kingdom by CPI Group (UK) Ltd, Croydon, CR0 4YY, UK

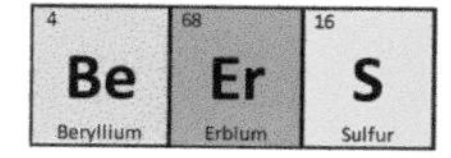

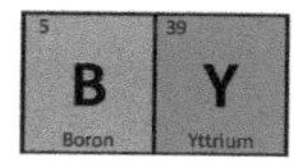

Dedication

This book is dedicated to Tim Watson, proprietor of the bijoux beer shop and bar 'Elementary' in Rhos-on-Sea, Conwy, North Wales who has plied the author with numerous fine beverages over the years (and where some of the more wacky ideas in this book probably came from). Long may it continue!

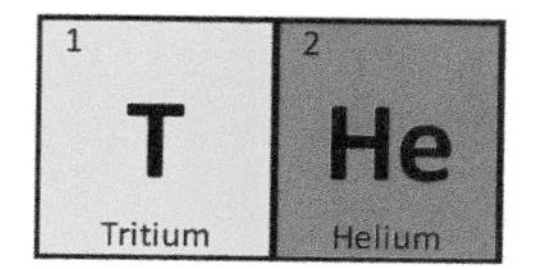

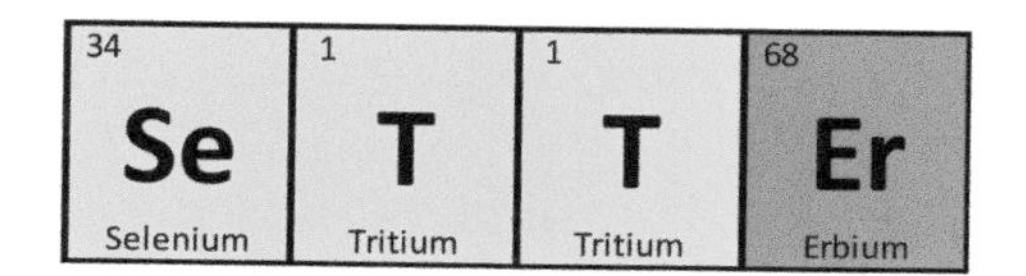

About the author

Paul Board is a scientist and committed cruciverbalist. As a 13 year-old, when Ilminster Boys' Grammar School was about to close after over 400 years, like other pupils he was allowed to take a couple of mementoes home. He chose a pair of boxing gloves and a Liebig Condenser. He never took up boxing, but he did graduate in chemistry at the University of Leeds in 1980 and has enjoyed a career in chemistry, including food analysis, environmental science and geochemistry both in the UK and overseas.

He has written for several scientific publications including *New Scientist*. For over a decade he has been compiling puzzles for the Royal Society of Chemistry for their magazines and online content (some of which made it into a book, *Chemistry Crosswords*, published by the RSC in 2017).

He is a Fellow of the Royal Society of Chemistry.

1	2		3	4	5	6	7	8	9	10	11	12	13	14	15	16	17	18
1 H																		2 He
3 Li	4 Be												5 B	6 C	7 N	8 O	9 F	10 Ne
11 Na	12 Mg												13 Al	14 Si	15 P	16 S	17 Cl	18 Ar
19 K	20 Ca		21 Sc	22 Ti	23 V	24 Cr	25 Mn	26 Fe	27 Co	28 Ni	29 Cu	30 Zn	31 Ga	32 Ge	33 As	34 Se	35 Br	36 Kr
37 Rb	38 Sr		39 Y	40 Zr	41 Nb	42 Mo	43 Tc	44 Ru	45 Rh	46 Pd	47 Ag	48 Cd	49 In	50 Sn	51 Sb	52 Te	53 I	54 Xe
55 Cs	56 Ba	57–70 La-Yb	71 Lu	72 Hf	73 Ta	74 W	75 Re	76 Os	77 Ir	78 Pt	79 Au	80 Hg	81 Tl	82 Pb	83 Bi	84 Po	85 At	86 Rn
87 Fr	88 Ra	89–102 Ac–No	103 Lr	104 Rf	105 Db	106 Sg	107 Bh	108 Hs	109 Mt	110 Ds	111 Rg	112 Cn	113 Nh	114 Fl	115 Mc	116 Lv	117 Ts	118 Og
			57 La	58 Ce	59 Pr	60 Nd	61 Pm	62 Sm	63 Eu	64 Gd	65 Tb	66 Dy	67 Ho	68 Er	69 Tm	70 Yb		
			89 Ac	90 Th	91 Pa	92 U	93 Np	94 Pu	95 Am	96 Cm	97 Bk	98 Cf	99 Es	100 Fm	101 Md	102 No		

Contents

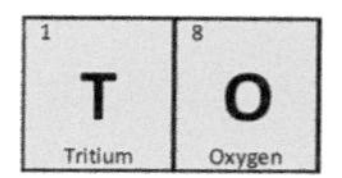

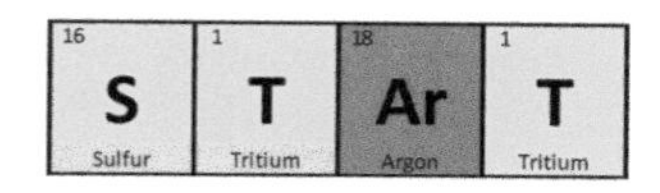

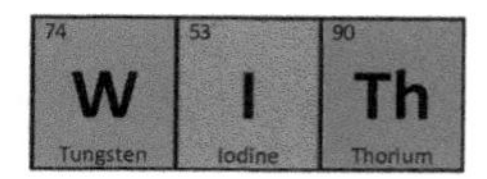

In October 2002, the Royal Society of Chemistry saw fit to award the much-loved and fictional sleuth Sherlock Holmes an Extraordinary Honorary Fellowship of the Royal Society of Chemistry, 100 years after the knighthood of his creator, Sir Arthur Conan Doyle (1859–1930).

As John Watson said at the ceremony "*Sherlock Holmes was way beyond his time in using chemistry and chemical sciences as a means of cracking crime.*"

But Holmes has also been misquoted on many an occasion: the immortal words, '*Elementary, my dear Watson*' never actually flowed from Conan Doyle's pen (although in 1988 Star Trek ran an episode entitled '*Elementary, Dear Data*'). Now, '*Elemental, my dear Watson*' would be more like it.

The ultimate representation of the elements takes the form of the Periodic Table, as seen in this book (and in chemistry laboratories and lecture rooms around the world), although hundreds of other Periodic Tables have since been proposed (some of them 3D). Of course, the table has been hijacked by many for other uses (see the excellent *The Science of Spices* by Dr Stuart Farrimond by way of example). The nuclear physicist Ernest Rutherford has been quoted (or misquoted) '*All science is either physics or stamp collecting.*' And when you look at the Periodic Table, appearing like so much philately, it maybe rings true.

The original Periodic Table really took shape in 1869 with Dmitri Mendeleev after a three-day card game of chemical patience, followed by a dream (realising he didn't have the whole deck, Mendeleev cleverly left blank spaces for predicted elements that on subsequent discovery proved the Russian chemist right). But the concepts of atoms and elements are hardly new. Back in 427 BCE, Plato's atoms took certain geometric shapes according to which elements they were, and there were only a supposed four elements at the time: fire (tetrahedron), air (octahedron), water (icosahedron) and earth (cube). By Shakespeare's time, only 14 elements were known, and it took an alchemist, Hennig Brand to discover the next one (phosphorus) in 1669.

The American mathematician and singer-songwriter Tom Lehrer cleverly kept his options open in the closing lines of his song *The Elements*. Recorded in 1959, it lists the 102 elements known at the time to the tune of The Major-General's Song from *The Pirates of Penzance*. We now have 118 but could there be more?

Whether or not we have ever stepped into a laboratory, the chemistry of the elements has impacted all our lives (maybe even saved it), yet chemistry sometimes gets a bad press. Way back in 1837, one of the greatest chemists of all time, Justus von Liebig (famous for his condenser) declared to delegates in Liverpool '*chemists are ashamed to be known by that name.*' This author is not one of them (and indeed once appeared in *The Sunday Times* brandishing a snazzy collection of Periodic Table neckties).

So, get your thinking caps on to tackle over 400 puzzles, covering subjects from the sub-atomic to the astronomic, brain teasers, botany and beasts, food and forensics, places, poisons and popular culture, scientists and superpowers.

A lot of fun and a bit of chemistry.

How to use this book

Questions and puzzles in each chapter are roughly ordered from easy through to difficult (or from talc-soft to diamond-hard, these being the opposite ends of the Mohs Scale of Hardness), although gauging the respective difficulties is a necessarily subjective rather than an objective process.

If you have the stamina to complete all the clues throughout this book which are specially marked with a ⚛, note the symbol of each element (those in the answers only, and several of the special questions have more than one elemental answer) in the Periodic Table (perhaps in pencil to begin with!) at the beginning of this book. (Note that there may be more than one question whose answer is a particular element).

However, there will be ten elemental symbols still left untouched in the Periodic Table after you've finished. When rearranged (and using each elemental symbol once and once only and not switching the letters in a two-letter elemental symbol) it will spell out the name of a scientist. For instance, if we had As, B, La, Ra, Si and U left over, we could make [La]U[Ra] B[As]Si, Laura Bassi (an Italian physicist and the first female professor of physics at any university). However, our ultimate answer was an American chemist credited with the invention of a very famous polymer. Who was that chemist and what was the polymer?

Note on nomenclature and spellings

For simplification and puzzling practicality, throughout these pages you will find the common rather than the official names of compounds, the latter as specified by IUPAC (International Union of Pure and Applied Chemistry), for instance 'acetone' rather than 'propan-2-one' although sometimes the common and IUPAC names are the same (such as benzene).

The Last Scientists (Image courtesy of the artist Georgia Wellman)

The Alchymist, In Search of the Philosopher's Stone, Discovers Phosphorus...by Joseph Wright of Derby

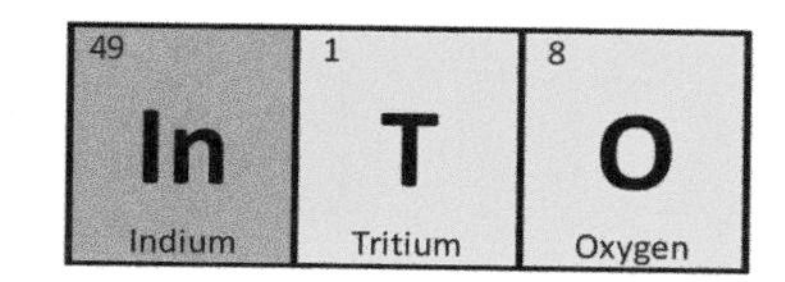

Alchemy (Lead into Gold)

...Alchemy is a pretty kind of game,

Somewhat like tricks o' the cards, to cheat a man

With charming.

Ben Jonson (1573–1637) in 'The Alchemist'

Chemistry has its origins in the archaic art of alchemy, both historically and etymologically. Indeed, the first element to be discovered since ancient times was discovered accidentally in 1669 by the German alchemist Hennig Brand when searching for the 'Philosopher's Stone', a substance which was believed to be able to transmute lead into gold. Brand discovered phosphorus and later sold the recipe (which involved boiling urine, so it was literally taking the P), so he was better off financially for his serendipity.

The Irishman Robert Boyle was born into a very rich family so had no need of such financial recompense. He is famously remembered for his book *The Sceptical Chymist* (and it was Boyle who was responsible for the dropping of the 'al' from alchemy).

The Philosopher's Stone did make someone else very rich more recently though, '*Harry Potter and the Philosopher's Stone*' being the debut novel of one J K Rowling.

1. *Chrysopoeia* was the 'science' of transmuting base metals into gold. Can you transmute Pb to Au in two simple steps, one letter at a time with this short word ladder (or step ladder), with each line being a normal word?

   L E A D

   _ _ _ _

   _ _ _ _

   G O L D

2. **Acrostic.** Solve the clues across in the left-hand grid, and reading down in the first column you will find the name of an alchemist. Transfer the letters in the left-hand grid into the right-hand grid as specified and you will find a quotation from him.

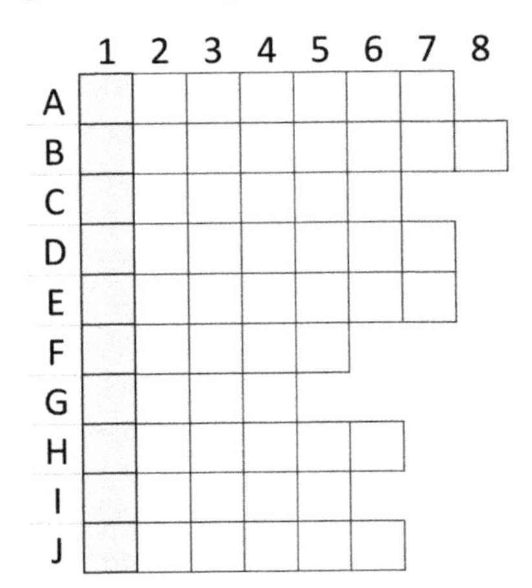

A7	B6		B5	A6	B2		D3	E3	H6
B3	C1	D1	I4	H1	D6	H4	E7	F1	
B1	B7	D7		B5	G3	F5	C6	G2	E4
E5	J4		D4	D5		B4	J1		
H6	E6	B3							
F5	C5	H2	I4	H1	D6	H4	H3	I5	G4
E2	I3	D6	H1	C4	D2	J4			

A. Alchemical apparatus named after a bird
B. The Triumphal (See E below) of _ _ _ _ _ _ _ _ by Basil Valentine
C. _ _ _ _ _ _ Boyle author of *The Sceptical Chymist*
D. First live album of *Dire Straits*
E. The Triumphal _ _ _ _ _ _ _ of (See B above) by Basil Valentine
F. Birthplace of alchemy (country)
G. Alchemists tried turning this into gold
H. Alchemists' celestial representation of the above answer
I. Fluid that the alchemist Henning Brand first isolated phosphorus from
J. Brimstone

3. Which very famous physicist, mathematician and President of the Royal Society (from 1703 to 1727) devoted a lot of his time to alchemy?

4. Which Danish alchemist and astronomer lost part of his nose in a duel at the age of 19, and had a replacement made out of a metallic element? And what is the element?

5. Geber (c721–c815), the Latinised name of *Jabir ibn Hayyan* was an Arabian alchemist. His name now forms the basis of a term relating to nonsense. What is it?

6. Link these seven metals to their 'heavenly bodies' (there's one obvious one!):

1. Copper	A Jupiter
2. Gold	B Mars
3. Iron	C Mercury
4. Lead	D Moon
5. Mercury	E Saturn
6. Silver	F Sun
7. Tin	G Venus

7. Metals also had ancient associations with the days of the week. Match these up:

1. Monday	A Copper, Cu
2. Tuesday	B Gold, Au
3. Wednesday	C Iron, Fe
4. Thursday	D Lead, Pb
5. Friday	E Mercury, Hg
6. Saturday	F Silver, Ag
7. Sunday	G Tin, Sn

8. Which wild plant was thought by alchemists to turn mercury into gold as the green leaves turn to a metallic blue-black tone when dried?

A. Plumbago
B. Dog's Mercury
C. Viper's Bugloss

9. Iron pyrite or iron sulfide, FeS_2 has another common name. What is it?

10. The early alchemist Mary the Jewess (also known as Mary the Prophetess) is credited with the invention of several chemical apparatus, including the one illustrated, which bears her name. A modified version is now used in kitchens. What is its name?

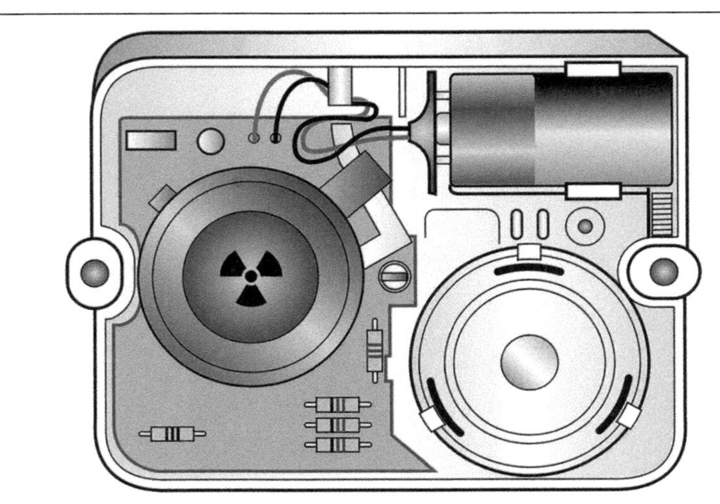

Smoke Alarm. Any smoke entering the radiation source chamber is ionised by the radiation, producing a current and thus setting off the alarm. ©Science Photo Library

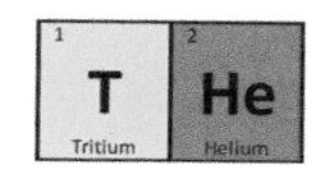

Science is built up of facts, as a house is with stones. But a collection of facts is no more a science than a house is a heap of stones.

Henri Poincaré (1854–1912)

You could be surprised by the chemistry and physics lurking in your home. From the chemical cornucopia hidden under the kitchen sink (and in your medicine cabinet), labour-saving devices to broken vase-saving glues, and the natural (and not so natural) radiation that could be a hazard (or could save your life). If you're lucky enough to have a garage or a garden shed (the latter probably treated with some wood preservative – see question 13), it's likely that many other molecular mixtures may be found lurking within.

You'll find some household names in some of the questions below, one or two having chemical connections in their brand names, albeit sometimes tenuous ones.

1. Which radioactive element was first produced in 1944 as part of the Manhattan Project but may now be found in many households inside smoke detectors (have you checked yours lately?), and what is the other element that it decays into?

2. Which colourless, odourless and tasteless gas produced by incomplete combustion is known as the *Silent Killer*?

Advert for Crane's Black Lead, ca. 1905

3. Black lead or plumbago is used to blacken and polish fire grates. What is the key ingredient?

4. Link these common household items with their active ingredients (although note there can be alternatives in some cases):

1. Antifreeze	A Acetone $(CH_3)_2CO$ (also known as propan-2-one)
2. Antacid	B Acetic acid CH_3COOH
3. Baking soda	C Sodium hypochlorite NaOCl
4. Bleach	D Ethylene glycol and/or propylene glycol
5. Drain cleaner	E Calcium carbonate $CaCO_3$
6. Nail varnish remover	F Sodium bicarbonate $NHCO_3$
7. Vinegar	G Sulfuric acid H_2SO_4

5. Which of the two above must never be used together in the home and why?

6. Which three of the compounds in the answers in Question 4 are not approved for food use in the UK?

7. Dependent on the geology of where you live (and which preventative measures you are taking), which radioactive noble gas may be permeating your home?

8. Which element's oxide coats the walls of self-cleaning ovens?

A. Cerium
B. Nickel
C. Copper

9. The glass in the windows of your home were possibly made by floating the molten glass on a molten element. Which element?

A. Mercury
B. Tin
C. Lead

10. The space between double-glazed windows is nowadays filled with which inert gas due to its poor heat conduction?

11. Which well-known brand of bath salts once claimed that it 'radiates oxygen (from where it got its name)?'

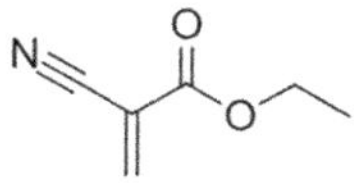

Ethyl cyanoacrylate

12. Superglue was invented in 1942 and was first available commercially in 1958. It uses cyanoacrylate compounds (such as the one above). These polymerise to make the adhesive in the presence of what compound?

13. Which well-known brand of wood preservative was invented in Denmark in 1911, originally for the preservation of fishing nets, but now is used extensively for the preservation and prettification of garden sheds and fences? Indeed, the company that makes it runs an annual Shed of the Year competition. The product's name derives from 'copper in oil.'

14. Which well-known brand of penetrating oil spray is composed of a mixture of hydrocarbons originally invented by the Rocket Chemical Company to protect the outer skin of the Atlas Missile, and what does its alphanumeric name stand for?

15. Which element is the active constituent of haemorrhoid cream but was also used by Victorian ladies as a skin whitener, sometimes with horrific results, as recounted by John Scoffern, surgeon and sometime assistant chemist at London Hospital in his *Chemistry No Mystery* (1839):

Hydrosulphuric acid gas enters into the composition of certain mineral waters; Harrowgate (sic) water, for instance, contains a large quantity of it; and connected with this subject, I have an anecdote to relate to you.

It was a practice with those ladies who were particularly ambitious of possessing a white skin, to daub themselves with a preparation of the metal _ _ _ _ _ _ _, which is one of these that sulphuretted hydrogen blackens. Now it is represented that on creditable authority, that a lady made beautifully white by this preparation, took a bath in the Harrowgate waters, when her fair skin changed in an instant to the most jetty black. You may judge how much was her surprise at this unlooked-for change; uttering a shriek, she is reported to have swooned; and her attendants, on viewing the extraordinary change, almost swooned too, but their fears in some measure subsided on observing that the blackness of the skin could be removed by soap and water.

© Macrovector/Shutterstock

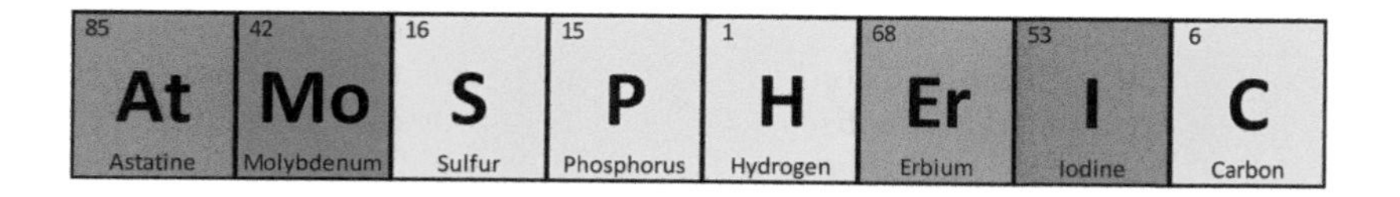

The word 'meteorology' (and 'meteor') derives from the Greek *meteoros*, meaning 'raised on high.' And it was a Greek, the philosopher and polymath Aristotle, who coined the phrase in his treatise *Meteorologica*, published around 340 BCE. Aristotle's concept of the Universe has one of the four elements, Earth at its centre with the other three, Water, Air and Fire radiating outwards respectively to the Moon, beyond which was the celestial sphere and the realm of 'astronomy' (for which see the 'Planets' section of this book). Of course, we now know that the Sun does not orbit the moon but Aristotle's *Meteorologica* (and much of his other work) influenced Western civilisation for millennia.

Nowadays, perhaps our commonest exposure to the science of meteorology is after the news with the weather report. But, of course, weather is not climate, and climate change due to anthropogenic altering of our atmosphere is now a collective concern.

1. At 78%, what is the main component of the Earth's atmosphere?

2. Which inert gas constitutes almost 1% of the Earth's atmosphere?

3. Link these celestial bodies with the major component of their atmospheres:

1. Earth, Pluto and Titan	A Carbon dioxide (CO_2)
2. Jupiter, Saturn and Uranus	B Hydrogen (H_2)
3. Mars and Venus	C Nitrogen (N_2)

4. Which of the above planets has clouds of sulfuric acid (H_2SO_4)?

5. Which element of our atmosphere is the name of a play written by biochemist Carl Djerassi (nicknamed 'the father of the pill') and chemist Roald Hoffmann (awarded the 1981 Nobel Prize in Chemistry)?

6. Inspired by sightings of the Brocken Spectre on the summit of Ben Nevis, the Scottish meteorologist Charles Thomson Rees Wilson (1869–1959) invented a device that was instrumental in the discovery of subatomic particles (including the positron, muon and kaon) and he was subsequently awarded the 1927 Nobel Prize in Physics. What was the device?

7. 'Every cloud has a silver lining' or so the saying goes. But some actually do (or rather did). *Operation Cumulus* was a cloud-seeding programme undertaken by the British military in the UK between 1949 and 1952 in an attempt to produce rain on order and thus bog down enemy tanks. This has been tenuously linked with the 1952 Lynmouth Flood Tragedy in Devon, UK (which claimed over 30 lives), although the link has never been proven.[†] What was the substance used?

A. Silver metal particles
B. Silver nitrate
C. Silver iodide

8. Which chemist, perhaps known more for his atomic theory, was also an avid meteorologist with his first publication (in 1793) being *Meteorological Observations and Essays* ?

A. Antoine Lavoisier
B. John Dalton
C. Henry Cavendish

9. Another chemist, Luke Howard (1772–1864) was also an amateur meteorologist who recorded London weather. Howard is referred to as 'The Namer of Clouds' on the blue plaque on his London home, and our cloud classifications today stem from his 1803 *Essay on the Modification of Clouds.* In 2018, which London football club paid tribute to Howard in its new stadium (close to where Howard lived) by naming viewing areas 'Stratus East' and 'Stratus West?'

A. Arsenal
B. Chelsea
C. Tottenham Hotspur

[†] The author's aunt Gwen and her two children John and Anne were rescued from the Lynmouth floods.

10. James Lovelock's invention of the Electron Capture Detector (ECD) in 1957 enabled him in the 1960s to be the first person to detect the widespread presence of man-made CFCs in our atmosphere. This data eventually led to the 1995 Nobel Prize in Chemistry being awarded to Paul Crutzen, Mario Molina and F. Sherwood Rowland *'for their work on atmospheric chemistry, particularly concerning the formation and decomposition of ozone.'* It was found that the manmade CFCs ('Freons') in fridges, spray cans and solvents were depleting the ozone layer once broken down by the higher UV radiation in the upper atmosphere (CFCs are being phased out under the Montreal Protocol). What does CFC stand for?

11. What causes acid rain?

12. Lightning strikes can dissociate the oxygen (O_2) molecules in the atmosphere to form what?

Cow wearing a ZELP (Zero Emissions Livestock Project) Harness to mitigate methane emissions (courtesy of zelp.co)

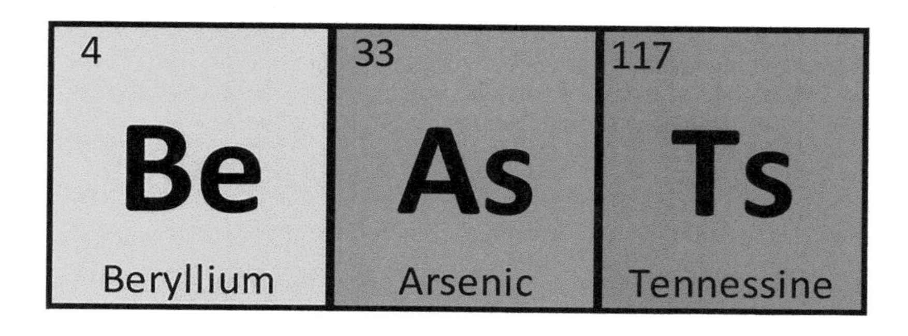

As an amateur naturalist (and a volunteer for the North Wales Wildlife Trust), the world of nature holds a special fascination for the author. As a puzzle compiler, it is not least the etymology (and sometimes the etymology of the entomology) of both the common names and the binomial names of animals that intrigues, which as you will see, like ants creeping into our kitchens also creep into chemistry and the wider fields of medicine.

Explore the animal kingdom in this section from vertebrate to invertebrate, wild to domesticated, pit viper to pet.

1. Add one letter to each of the following words and unscramble the resulting anagram to reveal an animal. Add the additional letter that was needed on the right and reading down you will find an alternative name for an element that gives rise to its elemental symbol. It also contains the name of two animals.
   - Lease ____________________ ____
   - Hearses ____________________ ____
   - Foaming ____________________ ____
   - Reciters ____________________ ____
   - Fog ____________________ ____
   - Mall ____________________ ____
   - Maestro ____________________ ____

2. The simplest organic acid is HCOOH, known as formic acid, so named after the Latin *formica* for the insect it was originally extracted from. What is the insect?

3. At the centre of haemoglobin, the metalloprotein that carries oxygen in our (oxygenated) red blood around our bodies, is an iron atom. Lobsters, snails and spiders have blue blood due to another metalloprotein, haemocyanin, that carries oxygen, at the centre of which is another element. What is that element?

4. Which (UK) permitted red food colouring is produced from crushed insects?

5. Schrödinger's cat is a famous feline which was used as a thought experiment. Link these other scientists to their respective animals:

1. Charles Darwin (1809–92)	A Bulldog
2. Albert Einstein (1879–1955)	B Beagle
3. Edwin Hubble (1889–1953)	C Bibo the parrot
4. Nikola Tesla (1856–1943)	D Diamond the dog
5. Thomas Henry Huxley (1825–95)	E Dolly the sheep
6. Sir Ian Wilmut (b 1944)	F Drooling dogs
7. Sir Isaac Newton (1642–1727)	G Macak the cat
8. Ivan Pavlov (1849–1936)	H Nicolaus Copernicus the cat

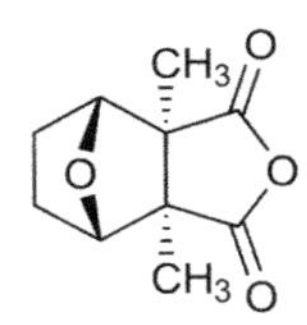

Reproduced from https://en.wikipedia.org/wiki/Cantharidin#/media/File:Cantharidin-2D.svg under the terms of the CC BY-SA license, https://creativecommons.org/licenses/by-sa/4.0/

6. Cantharidin (above) is a substance that is extracted from blister beetles, and historically was used as an aphrodisiac, sometimes leading to fatalities. What was it commonly known as?

A. Spanish beetle
B. Spanish fly
C. Ginseng

N	Na	I
Mo	S	Ar
C	B	Th

7. Using all 9 elemental symbols in the grid above (and *not* separating the letters of a two-letter symbol), spell out insects named after a red mineral form of mercury due to their distinctive red markings (2 words).

   Then answer these questions using some of the symbols, always including the middle symbol, and not switching letters with 2-letter symbols (using each symbol once and once only in each word):

   A. Narrows (5)
   B. Nerve agent (5)
   C. Agricultural buildings, plural (5)
   D. Disorderly crowds, plural (4)
   E. Wound (4)

Image courtesy of Mark Sheridan

8. Which yellow butterfly (above) is also named after an element, but an old name for that element?

9. Which noble gas has a popular aquarium fish named after it?

10. *Plumbago* was used to mark the sheep in the English county of Cumbria and was locally mined. What is this substance now known as?

A. Graphite
B. Lead chromate
C. Lead tetroxide

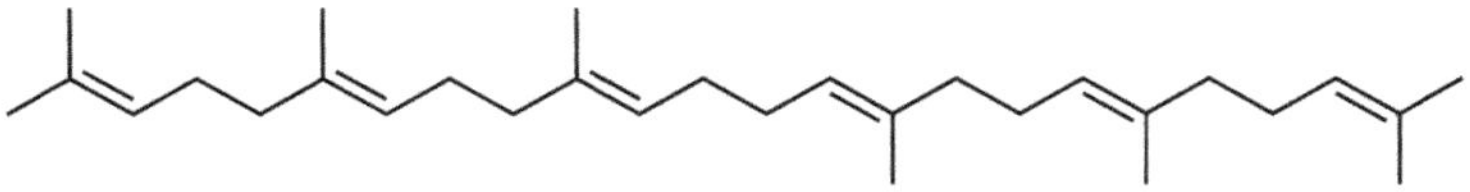

11. The above compound, squalene is used in cosmetics and vaccines. It was previously extracted from which of these animals:

A. Shark
B. Squacco heron
C. Whale

12. The German physical chemist and discoverer of the *Third Law of Thermodynamics*,[†] Walther Nernst (1864–1941) acquired a Prussian estate in 1920 with a thousand acres of land after selling his design for an electric lamp (known as the *Nernst glower*) for a million marks (he was also awarded the Nobel Prize in Chemistry the same year). On entering the cowshed he was surprised at how warm it was, assuming it was heated. But the heat was purely bovine and a result of metabolic processes. As a result, Nernst sold the cattle and invested in carp for his pond, saying that a thinking man cultivates animals that are in thermodynamic equilibrium with their surroundings and does not waste his money on heating the universe.[‡]

However, we are now aware of another issue. Methane, CH_4 is one of the greenhouse gases listed by the 1997 Kyoto Protocol and livestock production is the largest anthropogenic source. In June 2022, New Zealand proposed an Animal Gas Tax to start in 2025, taxing livestock belches. The British start-up company ZELP (Zero Emissions Livestock Project) uses catalytic cattle muzzle masks to convert the methane into carbon dioxide and water (see zelp.co. The device is currently undergoing field trials).

Roughly how much methane does a single cow produce on a daily basis (and 95% of it comes from the front end, not the back end)?

A. 20 litres
B. 200 litres
C. 2000 litres

[†] The Third Law of Thermodynamics concerns closed systems in thermodynamic equilibrium and states that the entropy of a system approaches a constant value when its temperature approaches absolute zero. Nernst's coworkers described his untidy study and laboratory as 'the state of maximum entropy.

[‡] This anecdote and many others may be found in Walter Gratzer's excellent '*Eurekas and Euphorias*' (see Bibliography).

13. Which pioneer of vaccination was elected to the Royal Society in 1788 for his study of the nesting habits of cuckoos?

A. Joseph Lister
B. Edward Jenner
C. Erasmus Darwin

14. It is said that bears with infected or broken teeth used to strip the bark of willow trees for its pain-killing properties. The active compound is salicylic acid (or its derivatives) from the Latin for willow, *Salix*. What common drug comes from it?

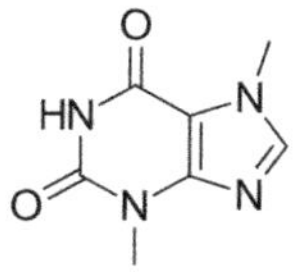

15. Chocolate is toxic to dogs due to the presence of the stimulant compound above. What is it?

A. Caffeine
B. Theobromine
C. Phenylethylamine
D. Oleic acid

16. Most of us have heard of the phrase 'red herring' but dead herrings start to glow after a while due to the slow burning of gases containing which element?

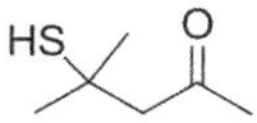

17. Cat ketone (4-mercapto-4-methylpentan-2-one or 4MMP for short, above) is one of the smelly substances in cat urine and is also found in which well-known wine grape?

A. Malbec
B. Riesling
C. Sauvignon blanc

18. Sea cucumbers are animals, not vegetables, belonging to the *echinoderms* (which include starfish and sea urchins). They have yellow blood due to the presence of which element?

A. Chromium
B. Manganese
C. Vanadium

19. *Captopril* was the first drug developed from an animal venom and is used for the treatment of hypertension and some types of congestive heart failure. Which animal's venom was used?

A. Brazilian pit viper
B. Geography cone snail
C. Emperor scorpion

20. Four riders were disqualified from jumping in the 2008 Olympics final since their horses tested positive for capsaicin. Capsaicin is used topically (*i.e.* on the skin) to soothe sore muscles but it is banned from the Olympics as it is also a hypersensitizing agent. Capsaicin is also the key component of which common spice?

21. What type of iron is named after a farmyard animal?

22. What is the colloquial name for the insects that defend themselves by spraying potential predators with a mix of chemicals (including two compounds used by the food industry for their flavour and aroma, the aldehydes *trans*-2-decanal and *trans*-2-octenal in one particular species)? In Taxco, Mexico, these insects are harvested and eaten (both alive and cooked) on Jumil Day (they are called *Jumiles* there), wrapped in tortillas and tacos.

23. What makes a gecko's feet stick so well to surfaces?

A. A natural glue
B. Van der Waals forces
C. Tiny hairs

24. A *sanguivore* is an animal which feeds exclusively on blood. One such animal has an anticoagulant called *draculin* in its saliva to stop the blood clotting on its way down. What is the animal?

25. Pheromones are airborne aphrodisiacs. *Bombykol* was the first pheromone to be characterised chemically and was discovered by the German organic chemist Adolf Butenandt (1903–95) in 1959. Bombykol gets its name from the Latin name of the insect it comes from, *Bombyx mori.* What is its common name?

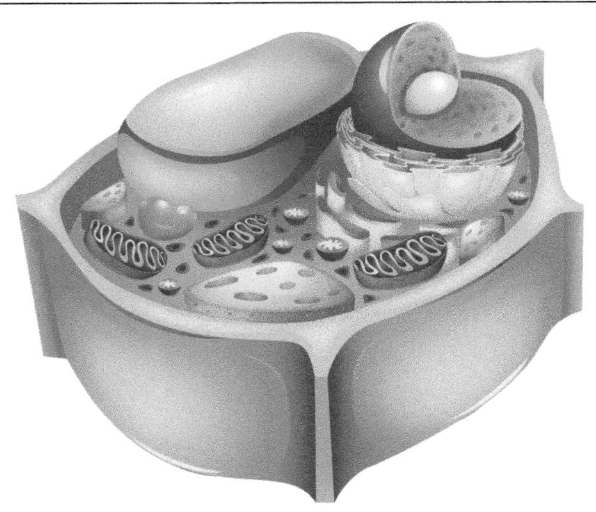

Schematic of a plant cell. Chloroplasts contain a pigment, chlorophyll, which is used in photosynthesis.
© BlueRingMedia/Shutterstock

The author recalls singing a folk song '*Botany Bay*' when at school, about the penal colony in New South Wales, Australia, but there are other Botany Bays scattered across the globe.

Pharmacology traces its beginnings to plants, plant lore and folk remedies, herbalists and the occasional quack.

Whether or not you've your own garden (or allotment), drop your anchor in this Botany Bay, put your feet up and enjoy a brew of biochemistry and botanicals (gin or no gin).

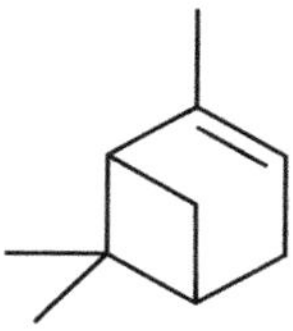

Reproduced from https://en.wikipedia.org/wiki/Pinene#/media/File:Alpha-pinen.svg, under the terms of the CC BY-SA 3.0 license, https://creativecommons.org/licenses/by-sa/3.0/.

1. The compound above (pinene) is the key flavouring component of a berry that is a must-have botanical for gin. Indeed, gin gets its name from the plant (as the Dutch called the alcoholic drink *genever* after the berry, and the term Dutch courage also derives from this drink). What is the English name for the berry?

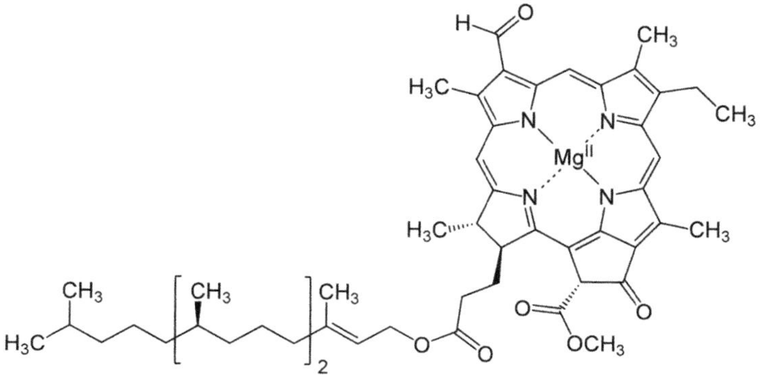

2. Magnesium is at the core of the chlorophyll molecule (above), the green pigment in plants essential for photosynthesis.

   Plants are of course the subject of interest to botanists, and Agnes Arber (1879–1960) was the first female botanist to be elected as a Fellow of the Royal Society (FRS). What else connects her to magnesium?

3. Link the plants with their pharmacologically active compounds:

1. Castor oil plant	A Thujone
2. Deadly nightshade	B Ricin
3. Hemlock	C Atropine
4. Peyote cactus	D Coniine
5. Wormwood	E Mescaline

4. What precursor to a common analgesic (painkiller) is named after the Latin name for the willow tree, the bark of which has been used since ancient times for the relief of pain and fevers, including by Hippocrates himself (known for his Hippocratic Oath)?

   A. Quercitin
   B. Salicylic acid
   C. Quinine

5. Physician, botanist, chemist and geologist William Withering (1741–1799) pioneered the use of the plant *Digitalis purpurea* for the treatment of heart complaints, and a carving of the plant is on his memorial stone at St Bartholomew's Church, Edgbaston. What is the plant's English name?

6. Which element was originally isolated from seaweed by the French chemist Bernard Courtois (1777–1838) after adding excess sulfuric acid to the seaweed ash when a cloud of purple vapour arose?

7. Similar to a word ladder, each answer below provides letters that must appear in the answer immediately following it. The puzzle's final answer is an English botanist, herbalist, astrologer and physician known for his *Complete Herbal*.

A. _ _ _	Playing card (3)
B. _ _ _ _	Delicate fabric (4)
C. _ _ _ _ _	Spear (5)
D. _ _ _ _ _ _ _	Hide (7)
E. _ _ _ _ _ _ _ _ _	Red food colouring (9)
F. _ _ _ _ _ _ _ _ _ _ _ _ _ _ _ _	Herbalist (8, 8)

8. During the First World War, children were asked to collect horse chestnuts ('conkers') and acorns, receiving 7s 6d per hundredweight (about 50 kg), as a source of starch. Was this for:

   A. Food uses
   B. Explosives
   C. Medicine

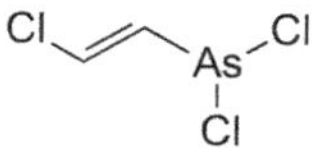

Reproduced from https://en.wikipedia.org/wiki/Lewisite#/media/File:Lewisite.svg, under the terms of the CC BY-SA 3.0 license, https://creativecommons.org/licenses/by-sa/3.0/.

9. If you can smell _ _ _ _ _ _ _ _ _ _ you're dead!

   This was the UK government advice during the Second World War, the missing word being a popular flowering garden and indoor plant, which apparently could smell like the organoarsenic blister agent and lung irritant known as *lewisite* (structure shown above) which it was feared could be dropped in bombing raids. What were the plants?

10. Oxalic acid is present in rhubarb, more so in the leaves, and is toxic at high doses. It proved fatal when children were given rhubarb leaves as a vegetable during the First World War. Oxalic acid pairs with the body's calcium to form insoluble calcium oxalate crystals, which accumulate in a particular organ of the body. What are these painful growths commonly known as?

11. The popular garden plant *Hydrangea* acts like a flowering litmus test,[†] turning blue in acidic soils and pink in soils with more lime in them (and some shades between). This colour variation is due to the plants' uptake (or absence) of which element?

    A. Aluminium
    B. Cobalt
    C. Copper

---

† Albeit a reverse litmus test. Litmus paper turns red with acid, blue with alkali. For more on the litmus test, see the Litmus Test section.

12. What makes stinging nettles sting?

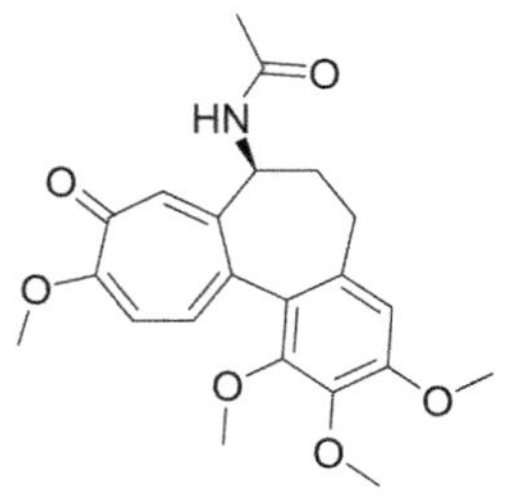

13. What flowering plant provides the spice saffron, and a toxic alkaloid colchicine (above), used as a treatment for gout?

14. Which flowering plant has a milky fluid in the unripe seed capsules containing morphine, codeine and papaverine?

15. Camphor and 1,8-cineole have repellent and insecticidal properties. They are naturally present in a garden plant that is sometimes put in wardrobes in its dried form to repel moths. What is the plant?

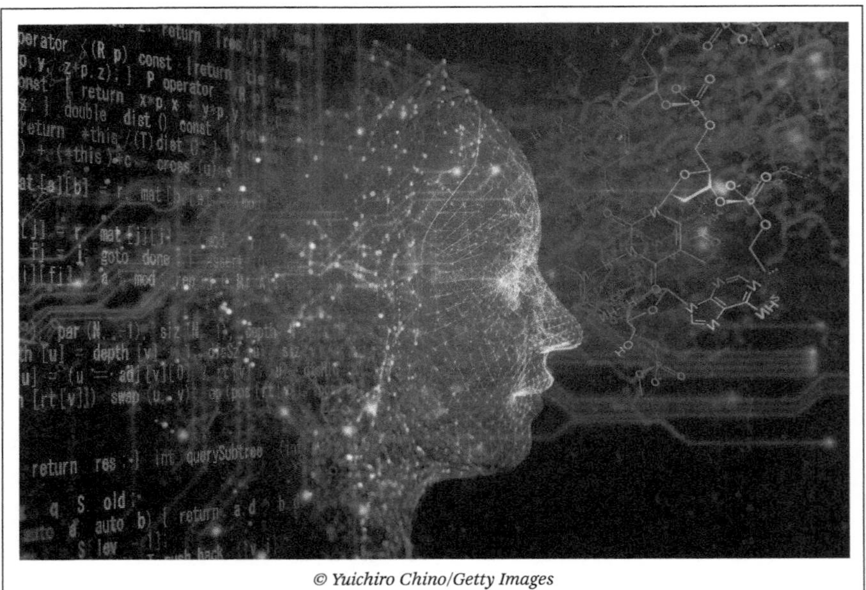

© *Yuichiro Chino/Getty Images*

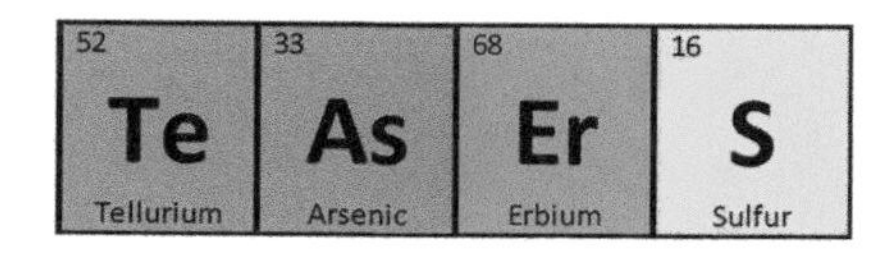

Lateral-thinking questions to periodically challenge the grey matter in your scientific brain. Cryptic clues, code-cracking chemistry, elementary puzzles (that are far from elementary), connection conundrums, fill the gap, spot the odd one out, and an A-to-Z scientific crossword puzzle (but with no clues as to where the answers should go).

1. Which metallic element can be transmuted into another metallic element simply by putting the word 'quick' in front of it?

2. Which salt may be found in barnacles, and which salt wrapped in sackcloth?

3. What connects the elements lithium, nitrogen, potassium and sulfur?

4. Caesium is the most electropositive stable element. What is the most negative?

5. Name the compound: DEFGHIJKLMNO (5, 5)

6. What connects the following metallic elements: antimony, iron, lead, gold, mercury, potassium, silver, sodium, tin, tungsten?

7. What do these compounds have in common: betaine, chitin, picric acid, psicaine, rhodamine, taurine?

8. What number comes next in this chemistry sequence and why? 9, 17, 35, 53, ?

9. Solve the following puzzle by filling in a letter in each cell (started off with S):

5	92	6	19	16 S
53		8		8
16 S	8	7	53	6
8		53		19
7	53	6	19	16 S

10. Each number in the row below corresponds to a letter. The letter is the same for each of the three answers A, B and C, which are all elements. Crack the code and once you have found all three elements, reading down the first letter of each answer will reveal another, bonus element.

1	2	3	4	5	6	7	8	9

A. 3711799217
B. 258217
C. 12345671

11. As above, this time with four codes to crack (the numbers stand for different letters than the previous question). Again, the first letter of each element reading down will reveal another element. What are the five elements?

1	2	3	4	5	6	7	8	9

- 912435167
- 1235
- 5835
- 482167

12. What connects the following elements: americium, curium, fermium, promethium, radon, samarium?

13. Fill in the blanks for these connected pairs of elements:

• Iron	Fermium
• Arsenic	_ _ _ _ _ _ _ _ (8)
• Copper	Curium
• _ _ _ _ _ _ _ (7)	Gadolinium
• _ _ _ _ _ _ _ _ _ (9)	Lawrencium
• Nickel	_ _ _ _ _ _ _ _ (8)
• Radium	_ _ _ _ _ (5)

14. And likewise for these (but with different connections):

• Carbon	Copernicium
• Chlorine	_ _ _ _ _ _ (6)
• _ _ _ _ _ _ _ _ (8)	Fermium
• Gold	_ _ _ _ _ _ _ _ _ _ (10)
• Fluorine	Iron
• Iron	_ _ _ _ _ _ _ (7)
• _ _ _ _ _ _ _ (7)	Tin

15. If Belgium is Bromine, Egypt is Calcium, Portugal is Lithium and Kenya is Sodium, what element is Norway?

16. All Greek to me. Crack these atomic code sequences to reveal the Greek letters:

A. 15, 16, 53
B. 53, 8, 73
C. 4, 73
D. 6, 1, 53
E. 45, 8

17. An A–Z of Science.[†] There are 27 clues to solve, each beginning with one of the 26 letters of the alphabet (and a bonus answer for the letter P). No clues as to where the answers should go but one letter has been provided to assist. Write your initial answers against the appropriate letters either side of the crossword until you're sure where to put the answer in the grid.

1	2		3		4			5	6		7		8	
							9							
10							11							
	12	C												
13											14		15	
16			17		18									
19	20				21				22		23			
24									25					
26											27			

A	N
B	O
C	P
D	Q
E	R
F	S
G	T
H	U
I	V
J	W
K	X
L	Y
M	Z

A. Italian scientist with a constant named after him (6, 8)
B. Pulsating (7)
C. Spherical bacteria (5)
D. Enemy of Dr Who (5)
E. Engrave with hydrofluoric acid (4)
F. Study of the unborn child (9)
G. Surgeons' workwear (5)
H. Water supply (7)
I. Demanding (9)
J. Geologic period (8)
K. Sycamore seeds (4)
L. Being (4)
M. Dye discovered by chemist William Perkin (5)
N. Atomic weapon (7, 6)
O. Light-sensitive protein found in the retina (5)
P. Excuse (7). The human soul, mind or spirit (6)
Q. Exploration (5)
R. Girl's name from the Greek for rose which gives its name to an element (5)
S. Flashes of light (14)
T. _ _ _ _ _ _ _ Pursuit, general knowledge board game (7)
U. Idealists (8)
V. Minecraft, Fortnite etc. (5, 5)
W. Tree source of salicylic acid (6)
X. Noble gas (5)
Y. Urania (10)
Z. Element used in galvanisation (4)

[†] Well, most of the answers are scientific!

18. Salt, lye and quicklime are common or old names for sodium chloride (NaCl), potassium hydroxide (KOH) and calcium oxide (CaO), respectively. Apart from previous names, what else do these three compounds have in common?

19. Fill the Gap. Place a well-known 3-letter word in each of the gaps to form six elements or compounds, and the shaded letters will form another element reading down.

I				I	U	M
R				I	U	M
C				I	N	E
A				I	N	E
			T	A	N	E
R				I	U	M

20. The three clues below each need a different two-letter elemental symbol put in front of the letters to make the names of an element or substance. Once solved, put the three elemental symbols together to spell out the name of another compound:

1. FFEINE, RBON, LOMEL
2. RINE, LENIUM, ROTONIN
3. DIUM, SULIN, DIGO

1		2		3	

21. And the same again:

1. MENE, PRITE, PREINE
2. RIUM, UXITE, RYTES
3. ON, URINE, OPRENE

1		2		3	

22. And one more:

1. TRE, OBIUM, CKEL
2. ETONE, TINIUM, RIDINE
3. ULIN, DOLE, OSITOL

1		2		3	

23. Neon is an inert gas, normally unwilling to chemically change due to its satisfaction with its full outer shell of electrons. However, neon can change in this word ladder puzzle. Change the word neon, one letter at a time (with the letter provided) to produce another word. What does neon change to in the end?

N	E	O	N	
				O
				M
				A
				L
				D
				E

24. Which of these elements is the odd one out (clue – America):

Aluminium, Argon, Calcium, Cobalt, Copper, Flerovium, Gallium, Indium, Lanthanum, Mendelevium, Manganese

25. What's significant about this sequence of elements (clue – empire):

Iodine, Vanadium, Xenon, Lithium, Carbon, Dysprosium, Mendelevium?

The International Prototype Kilogramme (IPK) or Le Grand K. See Questions 9 and 10. Reproduced from https://en.wikipedia.org/wiki/International_Prototype_of_the_Kilogram#/media/File:International_prototype_of_the_kilogram_aka_Le_Grand_K.jpg, *under the terms of the CC BY-SA 3.0 license,* https://creativecommons.org/licenses/by-sa/3.0/.

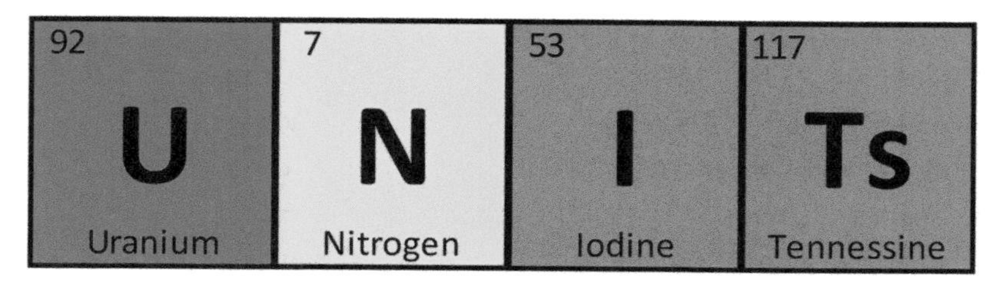

Calling All Units

The BIPM (Bureau International des Poids et Mesures) is an international organisation (currently with 59 member states) established by the Metre Convention in Paris on 20th May 1875, and is the home of the International System of Units (SI) and the International Reference Time Scale (UTC).

Quoting from the 2019 SI Brochure (which may be found at bipm.org):

> *The SI is a consistent system of units for use in all aspects of life, including international trade, manufacturing, security, health and safety, protection of the environment, and in the basic science that underpins all of these. The system of quantities underlying the SI and the equations relating them are based on the present description of nature and are familiar to all scientists, technologists and engineers... The definition of the SI units is established in terms of a set of seven defining constants. The complete system of units can be derived from the fixed values of these defining constants, expressed in the units of the SI.... Historically, SI units have been presented in terms of a set of – most recently seven – base units. All other units, described as derived units, are constructed as products of powers of the base units.*

1. List these masses in order of increasing magnitude:

A. Centigram
B. Femtogram
C. Kilogram
D. Microgram
E. Nanogram

2. Link these scientists to the SI Units named after them:

1. Alessandro Volta (1745–1827)	A. A Electric current
2. Anders Celsius (1701–1744)	B. Bq Radioactivity
3. André Marie Ampere (1775–1836)	C. C Electric charge
4. Charles Coulomb (1736–1806)	D. °C Temperature
5. Henri Becquerel (1852–1908)	E. Hz Frequency
6. Nikola Tesla (1856–1943)	F. T Magnetic flux density
7. Rudolph Heinrich Hertz (1857–1894)	G. V Electrical potential difference

3. What connects the SI Units of electrical resistance, units of conductance, and a scale of mineral hardness?

4. What element is used in the definition of the primary unit of time, the second, by the International System of Measurements (SI)?

   A. Caesium, Cs
   B. Gold, Au
   C. Hydrogen, H

5. The Curie (Ci) is a non-SI unit of radioactivity originally defined in 1910, based on the radioactive decay of an isotope of which element?

   A. Polonium, Po
   B. Radium, Ra
   C. Uranium, U

6. The newton (symbol N) is the SI unit of force, named of course after the physicist. Solve the clues below by changing one letter in each word (from the right column) and solving the anagram, starting with NEWTON and writing the new words below each other.

	N E W T O N	
• Urban dweller	_ _ _ _ _ _	I
• Spot	_ _ _ _ _ _	C
• Charged particle	_ _ _ _ _ _	A
• Type of energy	_ _ _ _ _ _	M
• Alga	_ _ _ _ _ _	D
• Territory	_ _ _ _ _ _	N
• Day of the week	_ _ _ _ _ _	Y

7. Which elements have the same symbols as the SI units for:

   A. Inductance
   B. Capacitance
   C. Force
   D. Electrical potential difference

8. The elemental symbol for gold is Au, derived from the Latin name *aurum*. AU is also a scientific unit. What does it stand for?

9. From 1889 to 2019, the international standard kilogram was a metal cylinder known as the IPK (for International Prototype of the Kilogram) or Le Grand K, stored at the BIPM on the outskirts of Paris (and pictured in the opening page of this section). However, since it was† losing mass, the kilogram was redefined based on physical constants in 2019. 10% of the metal in the cylinder was iridium. The other 90% was a precious metal. Which precious metal?

10. The kilogram is one of the seven SI base units. What are the other six?

---

† Ever so slowly.

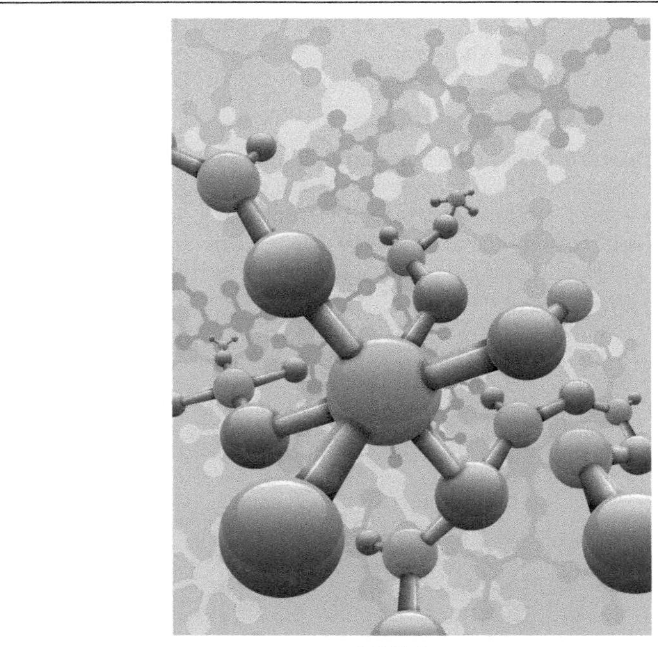

© Getty Images

Compound 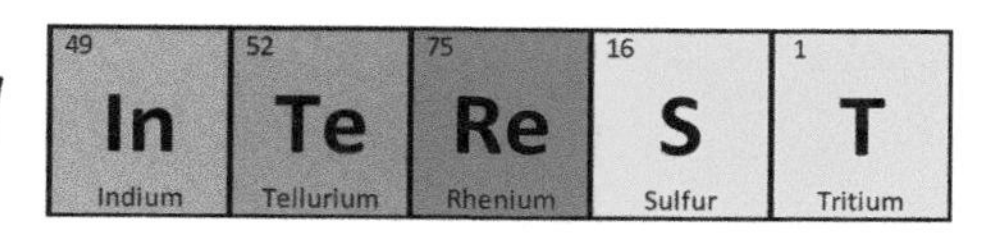

By way of introduction to the magical world of molecules, here's a little ditty by the author about an important building block of many organic molecules, benzene (with apologies to William Blake):

Benzene! Benzene! Burning bright
Belching engines day and night
What immortal hand or eye
Could frame Kekulé's symmetry?
Who'd have thought your Carbon Six
Could have produced such toxic tricks
Or provide the building blocks
For a plastic world (and cure the pox)?

Aesthetic, perfect aromatic
Substitutes produce chromatic
Dyes that brighten every day.
Thank you Mr Faraday.
Clothe our backs and cure our ills
Blow or dull our brains with pills
Ironic that your homologues
Pollute our land and stock our smogs
Benzene – your hydroxyl daughters
Need locking up, they pollute our waters
Adding chlorine provides persistence
(Target organs keep your distance).

Doubt Kekulé ever dreamt
Of such riches (or torment).
Oh benzene whether bound or free
Did He who made the Lamb make Thee?
Benzene! Benzene! Burning bright
Belching engines day and night
What immortal hand or eye
Could frame Kekulé's symmetry?

Explanatory notes:

(1) *Benzene is a constituent of petrol.*

(2) *Early gas lighting relied on the presence of traces of benzene which burned with a bright flame. Gas mantles shone brightly when heated. They were made of thorium oxide plus 1% cerium oxide. (The cerium was added to produce a softer glow. It also acted as a catalyst ensuring more complete combustion of the gas).*

(3) *Friedrich August Kekulé (1829–96) is attributed with elucidating the symmetrical ring structure of benzene* C_6H_6, *claiming as an old man that the idea of the carbon chain had first occurred to him in the summer of 1854 on top of a London omnibus, although the matter remains controversial.*

Notwithstanding that, A. W. Hofmann (1818–1892), German organic chemist, said in a speech at the Kekulé 'Benzolfest', 1890:

> *"Organic chemistry before Kekulé spread his wings was like a merrily splashing torrent; there were so many stones in the water that one could still cross it without getting wet. Today, the torrent has become a deep and massive stream; the eye can hardly see the opposite bank, and proud, richly loaded fleets rock gently on its broad surface.*
>
> *With the concept of the benzene ring the number of organic compounds all at once seems, one might almost say, to have increased to infinity. In the benzene nucleus we have been given a soil out of which we can see with surprise the already-known realm of organic chemistry multiply, not once or twice but three, four or six times just like an equivalent number of trees. What an amount of work had suddenly become necessary, and how quickly busy hands found to carry it out! First the eye moves up the six stems opening out from the tremendous benzene trunk. But already the branches of neighbouring stems have become intertwined, and a canopy of leaves has developed which becomes more spacious as the giant soars upwards into the air. The top of the tree rises into the clouds where the eye cannot yet follow it. And to what extent is this wonderful benzene tree thronged with blossoms! Everywhere in the sea of leaves one can spy the slender hydroxyl bud; hardly rarer is the forked blossom which we call the amine group, the most frequent is the beautiful crossed-shaped blossom we call the methyl group. And inside this embellishment of blossoms, what a richness of fruit, some of them shining in a wonderful blaze of colour, others giving off an almost overwhelming fragrance! Understandably, there is also no dearth of industrious workers busily striving to collect the harvest. Keen climbers have already clambered up to the third or fourth branch; some of them we can see working at a dizzy height. Most of them, however, are in the bottom branches of the benzene tree. Some have collected enough and are about to get down, others still cannot separate themselves from the rich harvest, and yet others are quarreling with their neighbours about the harvest..."*

(4) *Many synthetic dyes are based on substituted benzene molecules such as aniline.*

(5) *Michael Faraday's portrait used to be found on the back of an old UK twenty-pound note. He was replaced by Edward Elgar, who, interestingly, also practised chemistry as a pastime.*

(6) *Benzene was used in the production of artificial leather.*

(7) *Benzene is used in the manufacture of medicinal chemicals.*

(8) *Benzene is a known carcinogen and can pollute both the air we breathe and the land we live on.*

(9) *Phenols (benzene substituted with –OH groups) and related compounds are particularly hydrophilic (water loving) and can pollute our waterways.*

(10) *Organochlorines such as organochlorine pesticides (DDT is an example) and polychlorinated biphenyls (PCBs) do not readily break down in the environment and tend to bioaccumulate due to their lipophilicity (affinity for fat), causing problems particularly at the higher ends of the food chain.*

1. See if you can discover the six compounds and classes of compounds here (not the formulae but the actual words, some old, some new). For instance, Her f[ace tone]d gave B[eth a ne]w countenance contains 'acetone' and 'ethane.'

Ask whether I'm in England. Sure am! But a neighbour's in a bad spot as he's being pestered.

2. 'In the limelight' is a phrase that originally derives from the use of which compound to produce a brilliant white light?

A. Ascorbic acid
B. Citric acid
C. Calcium oxide

3. Which compound is used in some countries as an electoral stain that is applied to the forefinger of voters to prevent multiple voting by the same person?

A. Cobalt blue
B. Potassium permanganate
C. Silver nitrate

4. DHMO, dihydrogen monoxide, a chemical that has killed millions but is also the stuff of life. What is it normally known as?

5. Compounded compounds. The following six compounds consist of two words, but they've been jumbled up. To make it slightly easier than an anagram, despite the letters being jumbled, letters are in the right order as they appear in the two words.

For instance: S D I O I L I X I C O N D E reveals SILICON DIOXIDE (SiO_2) as follows: [S]DIO[I][L][I]XI[C][O][N]DE

A. C A C A L R C I B U O N A T M E
B. N O X I T R O U I D E S
C. S I N I L V T R A E T R E
D. C A T R E T B R A C O H L O R N I D E
E. C O S U P L F A P E T E R
F. S C H O L D I O R U I D E M

6. Fill in the missing cells to produce compound formulae and elemental isotopes (the latter, for the purposes of puzzling practicalities, allowing the elemental symbol before or after the numbers denoting the number of neutrons in the isotope) Using the following numbers and letters (letters and numbers being allowed in the grid more than once):

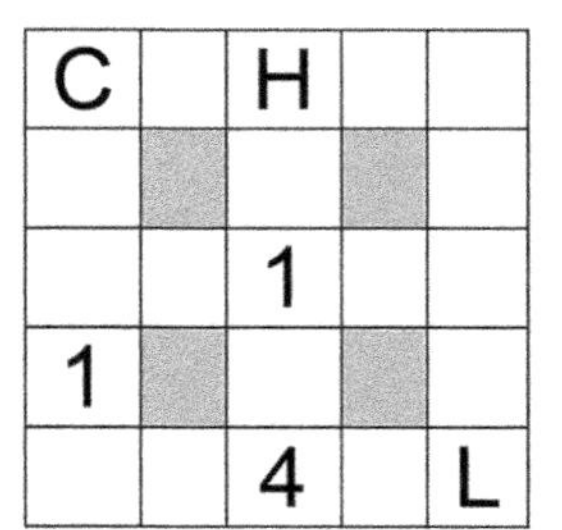

1	2
4	5
7	8
9	C
F	G
H	L
N	

A few extra hints:

- Ammonium chloride
- Pentane
- Pyridine

7. Complete this crossword combining compound formulae, atomic isotopes, elemental information and popular culture. The yellow cells contain a number rather than a letter. Again, this puzzle allows the elemental symbol before or after the numbers denoting the number of neutrons in the isotopes.

1		2			3	4			5	6		7
				8				9				
		10						11				
12				13								
												14
15		16		17			18					
						19						
				20				21		22		
23												
24						25						
26					27				28			

Across

1 Ethene (or ethylene) (4)
3 Rock group from Lincoln Park, Michigan, first album *Kick out the Jams* (3)
5 Registration number of the Batmobile from the 1966–68 Batman TV Series (4)
10 Red lead (5)
11 Canadian rock band, first album *All Killer, No Filler* (5)
12 Atomic number of the first of the superheavy elements (3)
13 Most abundant isotope of the metallic element whose name comes from the Greek for 'smell' (5)
15 Type of fatty acid found in fish oils, phytoplankton and marine algae, which plays an important role in human metabolism (7)
18 The postcode of BBC's Broadcasting House in London, the first half of which became the title of a mockumentary sitcom (6)
20 Least abundant natural isotope of the metallic element named after the Greek goddess of the rainbow, due to the colours of its salts (5)
22 Most common isotope of the element of diamond, graphite and buckyballs (3)
24 Track on Garth Brooks' album *Gunslinger* (5)
25 Mercuric chloride, once known as corrosive sublimate (5)
26 Most abundant isotope of a halogen that was used as a poisonous gas in World War 1 (4)
27 Hydrogen sulfide (3)
28 Birthday of physicist Edwin Herbert Land (1909–1991) who invented the Polaroid camera (4)

Down

1 Amyl nitrite (8)
2 Phosphoric acid (5)
4 Anthracene (6)
6 The licence plate of Bruce Almighty (played by Jim Carrey) in the 2003 US comedy film of the same name (7)
7 Synthetic isotope of tungsten (4)
8 Humanoid robot character in Star Wars (4)
9 Electron configuration of lithium (6)
14 Sodium dichromate (8)
16 US rap rock band which appeared on The X Factor and subsequently signed up with Simon Cowell's Syco Records (7)
17 1992 US science fiction horror film starring Sigourney Weaver as Ellen Ripley (6)
19 Chrysene (6)
21 English rock band named after a volumetric measurement (4)
22 Least abundant of the stable barium isotopes (5)
23 Isotope of technetium with a half-life of 4.2 million years (4)

8. See if you can link the old names with the new:

Old	New
1. Aqua regia	**A** Ammonia
2. Aqua fortis	**B** Ammonium sodium phosphate
3. Blue vitriol	**C** Antimony potassium tartrate
4. Corrosive sublimate	**D** Copper sulfate
5. Lunar caustic	**E** Hydrochloric acid
6. Microcosmic salt	**F** Hydrogen cyanide
7. Muriatic acid	**G** Mercuric chloride
8. Prussic acid	**H** Mixture of nitric acid and hydrochloric acid
9. Saltpetre	**I** Nitric acid
10. Spirits of hartshorn	**J** Potassium nitrate
11. Tartar emetic	**K** Silver nitrate

9. The following pairs of molecular formulae are linked by the fact that the name of the second (not necessarily the official IUPAC name) may be found in the former just by taking out one or more of the letters (without mixing those letters up). For instance M(ethane) and Ethane come from the formulae CH_4, C_2H_6. Name the compounds:

A. $C_6H_{12}O_6$, $C_{12}H_{22}O_{11}$
B. C_9H_7N, $C_{20}H_{24}N_2O_2$
C. $(CH_2)_4(NH)_2$, $C_5H_4N_4$
D. $(CH_2{=}CH)_2$, C_4H_{10}

10. The following partial words may be converted into compounds or groups of compounds by adding the same elemental symbol front and back throughout. What is the element?

_ _ UL _ _
_ _ SUL _ _
_ _ DIGOT _ _
_ _ DIRUB _ _
_ _ TERLEUK _ _

11. Which compound shares its popular name with a town in Surrey that holds a famous horse race?

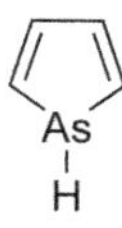

Reproduced from https://en.wikipedia.org/wiki/Arsole#/media/File:Arsole.png, under the terms of the CC BY-SA 3.0 license, https://creativecommons.org/licenses/by-sa/3.0/.

12. The organoarsenic compound above (C_4H_4AsH) has a rather unfortunate common name. What is it?

13. Here are some more compounds with funny (although not necessarily their IUPAC) names. Which is the odd one out?

Olympicene

Penguinone

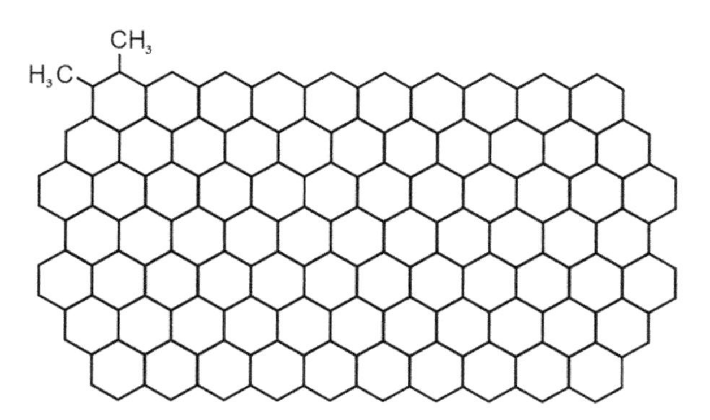

1,2-Dimethyl chickenwire

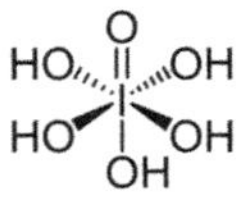

Periodic acid

14. Link these molecular structures with the compound names:

1

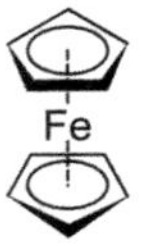

2

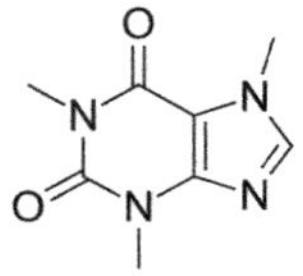

3

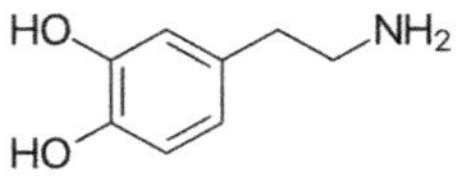

4

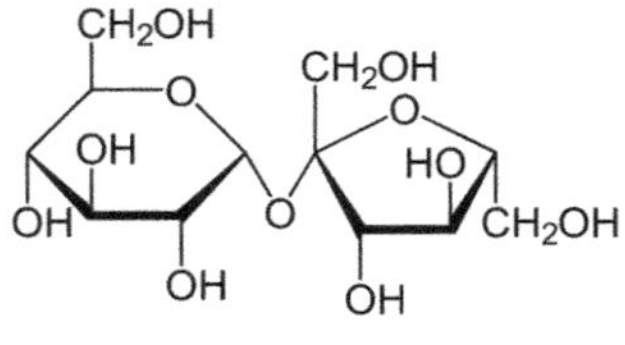

5

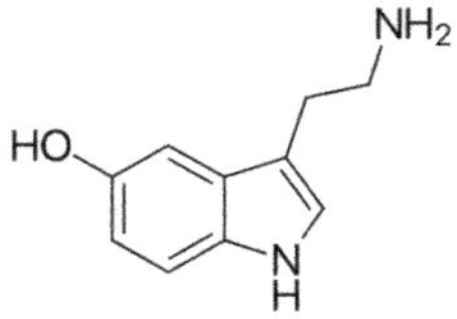

6

a. Caffeine
b. Dopamine
c. Ferrocene
d. Octane
e. Serotonin
f. Sucrose

15. Similar to a word ladder, each answer below provides letters that must appear in the answer immediately following it (but not necessarily in the same order). The puzzle's final answer is a compound.

A.	_ _ _	Jogged (3)
B.	_ _ _ _	Precipitation (4)
C.	_ _ _ _ _	Coach (5)
D.	_ _ _ _ _ _ _	Toilet (7)
E.	_ _ _ _ _ _ _ _	Pause (8)
F.	_ _ _ _ _ _ _ _ _ _	Unimportant (10)
G.	_ _ _ _ _ _ _ _ _ _ _ _ _	Compound (6, 7)

16. This is a representation of a crystal lattice structure called body-centred cubic (or bcc),[†] as it may be found in caesium chloride, for example:

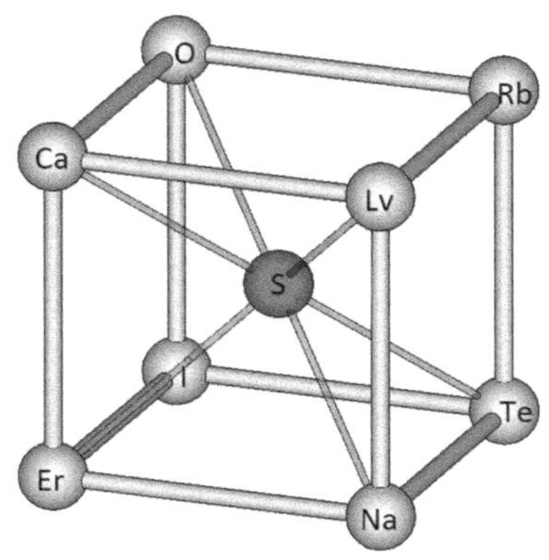

Adapted from https://en.wikipedia.org/wiki/Cubic_crystal_system#/media/File:CsCl_crystal.png, under the terms of the CC BY-SA 3.0 license, https://creativecommons.org/licenses/by-sa/3.0/.

Using all 9 elemental symbols in the structure above (and *not* separating the letters of a two-letter symbol), spell out the name of a real chemical compound (two words).

Then answer these questions using some of the symbols always including the middle symbol and not switching letters with 2-letter symbols (using each symbol once and once only in each word):

A. Greek god of love (4)
B. Location (4)
C. Globes (4)
D. Guide (5)
E. Someone like you? (6)

17. The reaction below shows the formation of a common compound. What is its everyday four-letter name?

$$4Fe(s) + 3O_2(g) \rightarrow 2Fe_2O_3(s)$$

[†]This is just using the structure for puzzling purposes and not an actual compound.

18. The author's full name is Paul Christopher Board, or P. C. Board for short. As a youngster he recalls seeing 'P C Boards' (Printed Circuit Boards) for sale in electronics magazines. Even shorter, PCBs are a class of manmade chemicals that have unfortunately caused much environmental damage, accumulating up the food chain.

Some examples are given below:

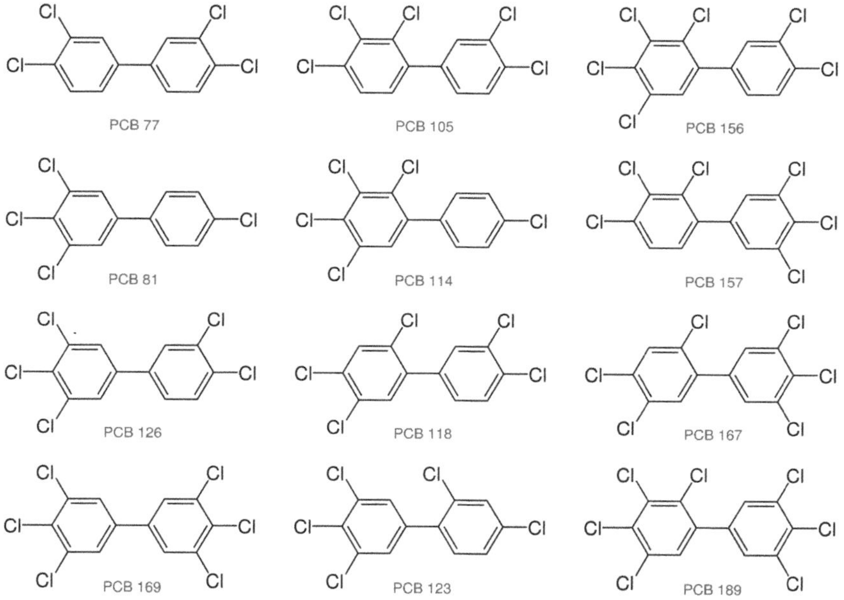

PCB 77 PCB 105 PCB 156

PCB 81 PCB 114 PCB 157

PCB 126 PCB 118 PCB 167

PCB 169 PCB 123 PCB 189

What does PCB stand for?

19. Acrostic. Solve the clues about compounds and minerals across in the left-hand grid. Reading down in the first column you will find another compound. Transfer the letters in the left-hand grid into the right-hand grid as specified and you will find an appropriate quotation and the name of the scientist who wrote it.

	1	2	3	4	5	6	7	8	9	10
A										
B										
C										
D										
E										
F										
G										
H										
I										
J										

A1	G3		B4	C4	A6	C3	J5	B5	F7	F3	E8
A7	E1	E9		B4	C4	I2	I6	C1		E7	E3
B4	J6	I7	G1	I4	G3		G3	C2			
I3	B3	A2	J2	A4	C4	F5	J5	G3			
G1	D3	G7		I8	E7	D6	D7	J5	H4	D5	E8
B4	D9		B4	C4	G6	C3	H2		C4	J2	
E7	C4	D8		C2	E4		F1	G4	E7	A3	E8
E3	D10	H1	J5	J1	F2	E5	G1	I4			
							G3	E7	J5	J5	G2

A. NH_3
B. Columbite

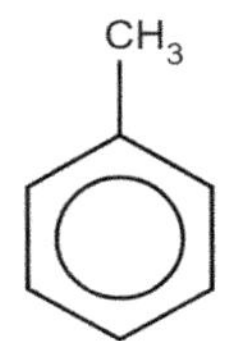

C. Compound above
D. D_2O (2 words: 5, 5)
E. Trisaccharide found in beans
F. Amino acid
G. Amino acid
H. Alkane
I. Niccolite
J. Medication obtained from the plant *Ephedra sinica*

20. This puzzle is based on the structure of a polycyclic aromatic hydrocarbon (PAH), hexabenzocoronene $C_{42}H_{18}$. There are six questions with six 3 letter answers, written clockwise or anticlockwise but starting in any of the 3 hexagons associated with each question number. Once complete, the inner circle of hexagons (without the double bonds) will reveal a 6-letter element when reading clockwise. The six outer hexagons will also reveal another element, reading anticlockwise. (The inner benzene ring remains empty). One of the 3 letter answers is provided by way of example.

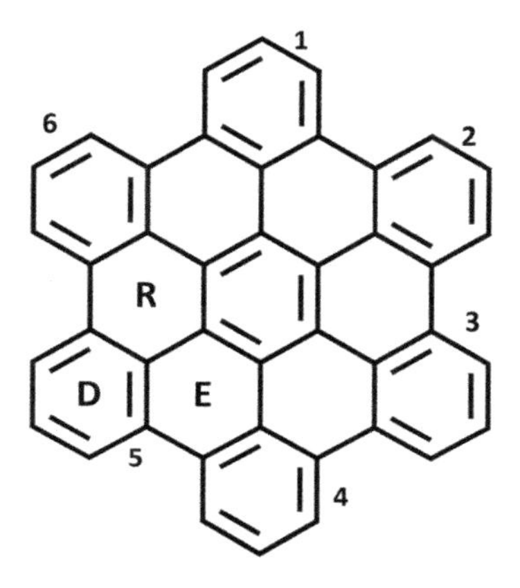

1. Legendary bird of prey
2. Cleaning implement
3. Young dog
4. Pastry
5. Colour of cochineal
6. Vehicle

21. This quick puzzle is again based on a PAH, coronene, $C_{24}H_{12}$. Rearrange the elemental symbols below to spell out the name of another PAH:

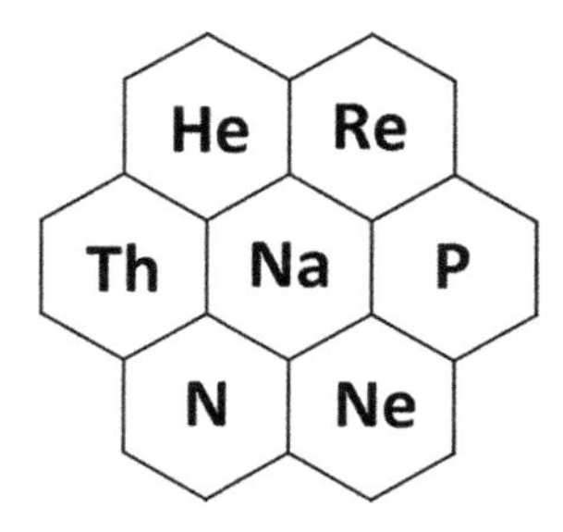

22. This puzzle is also based on coronene. Using the elemental symbols (and not switching the 2 letters in each symbol), place each of the seven elemental symbols in the hexagons to spell out three 6-letter words reading diagonally.

In each of the four puzzles there will be a shared element in the middle hexagon which when combined and reading across will spell out the name of another compound.

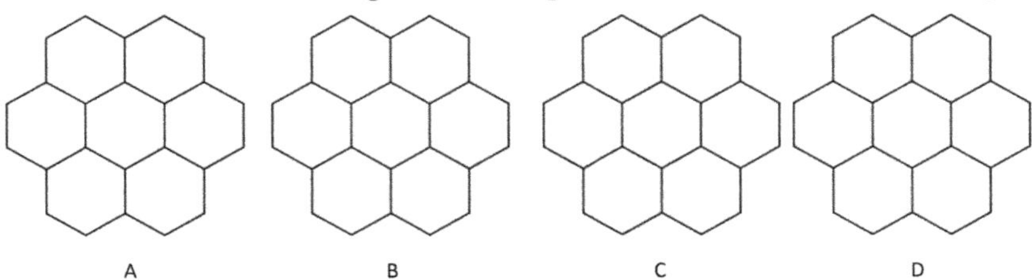

Elements:

A. Ar, Ba, Cs, Er, Ho, Ne, Si
B. Ca, Ce, Co, Cs, Li, Po, Re
C. Bi, Co, In, Nd, Re, Rn, Se
D. Ar, Ge, Li, Ne, Re, Te, Ts

23. And the same again. This time the four shared elements will spell out the name of an ore.

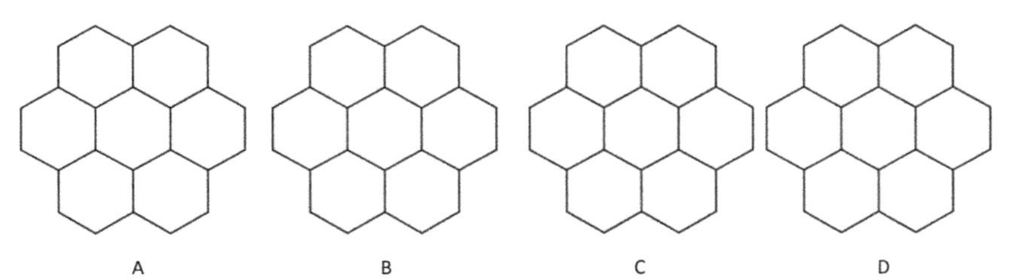

Elements:

A. Al, Er, Es, Fe, Ho, Li, Ne
B. Al, Mo, Nd, Os, Re, Se, Te
C. Al, Ca, Co, Cs, Ge, Ne, Ni
D. As, Ce, In, Nd, Ne, Rn, Te

24. In organic chemistry, the terms *ortho-*, *meta-* and *para-* (or *o-*, *m-* and *p-*) relate to positions on the benzene ring:

CH_3

benzene 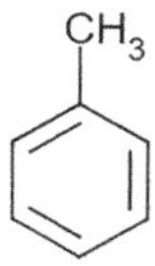 toluene

CH_3 CH_3 CH_3 CH_3

CH_3

CH_3

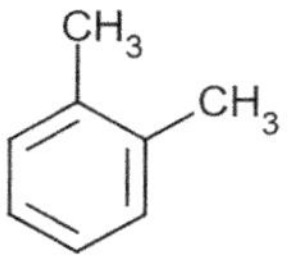

ortho-xylene 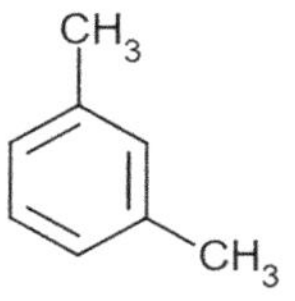 *meta*-xylene  *para*-xylene

So what would you call these compounds:

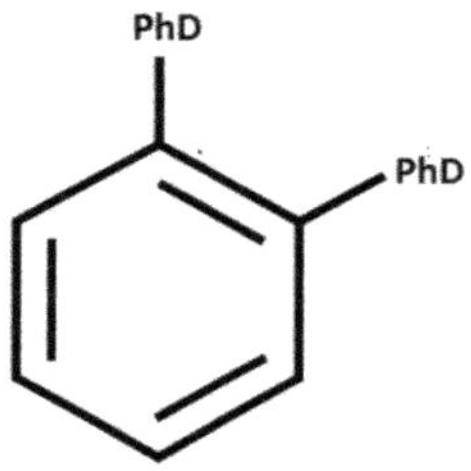

PhD

PhD

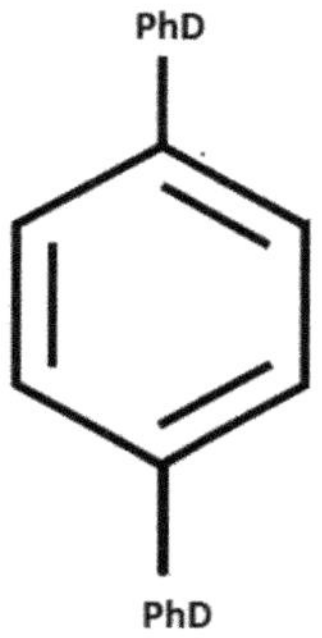

PhD

PhD

PhD

25. The organic compound below has a name that that sounds the same as a chemical element. What is that element?

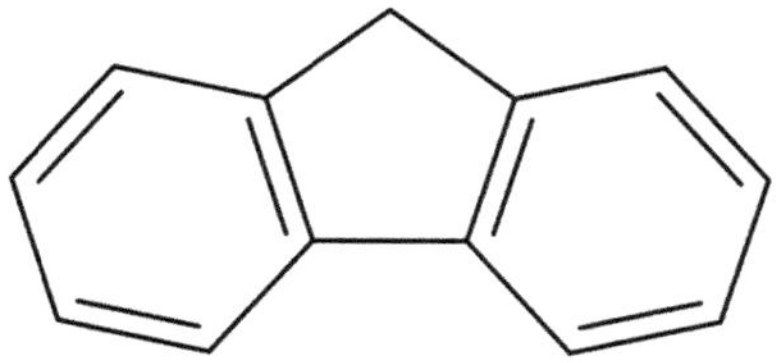

26. On 4 August 2020, thousands of tons of ammonium nitrate (NH_4NO_3) exploded in a warehouse in Beirut, killing over 200 people and injuring more than 5,000. What is this compound's main use?

27. For this puzzle, insert the six shapes (based on polycyclic aromatic hydrocarbons, PAHs) into the honeycomb grid above them so that the elemental symbols in each hexagon, reading left to right, spell out the surnames of seven scientists. No need to rotate any of the shapes. Who are the scientists?

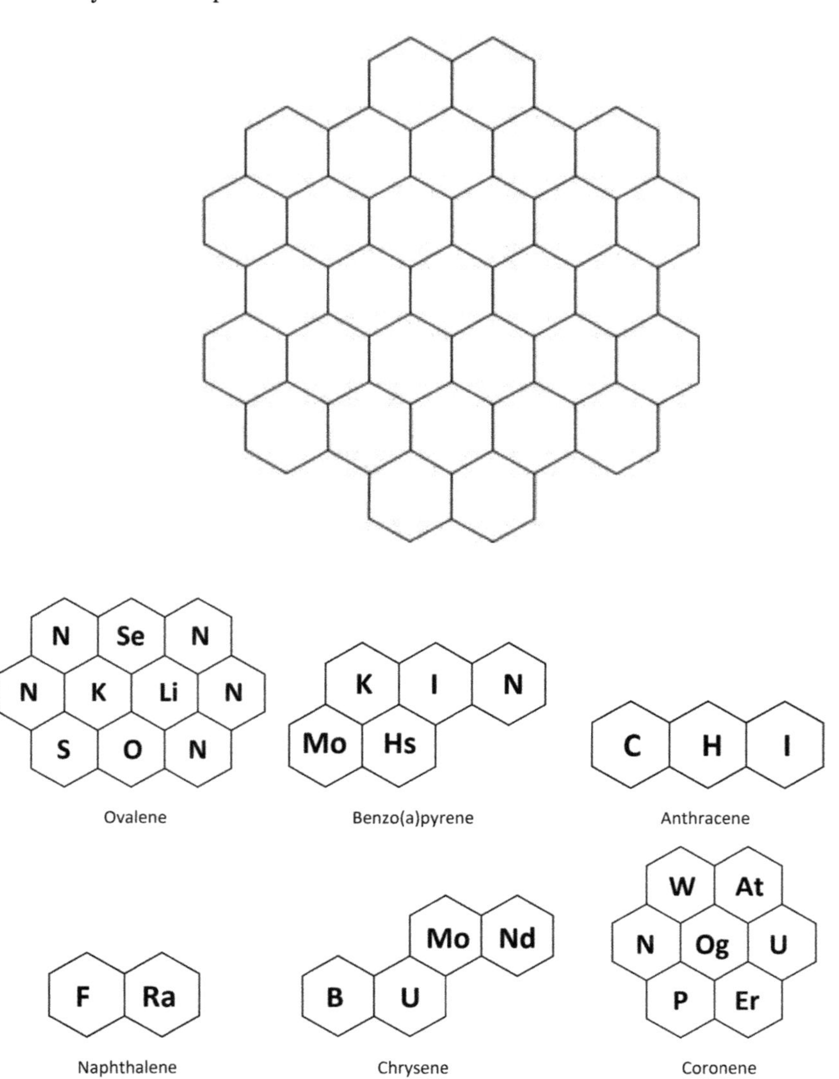

28. Similarly for this puzzle, insert seven shapes (again based on PAHs) into the honeycomb grid above them so that the elemental symbols in each hexagon reading left to right spell out seven forms or sources of energy. No need to rotate any of the shapes. What are the words?

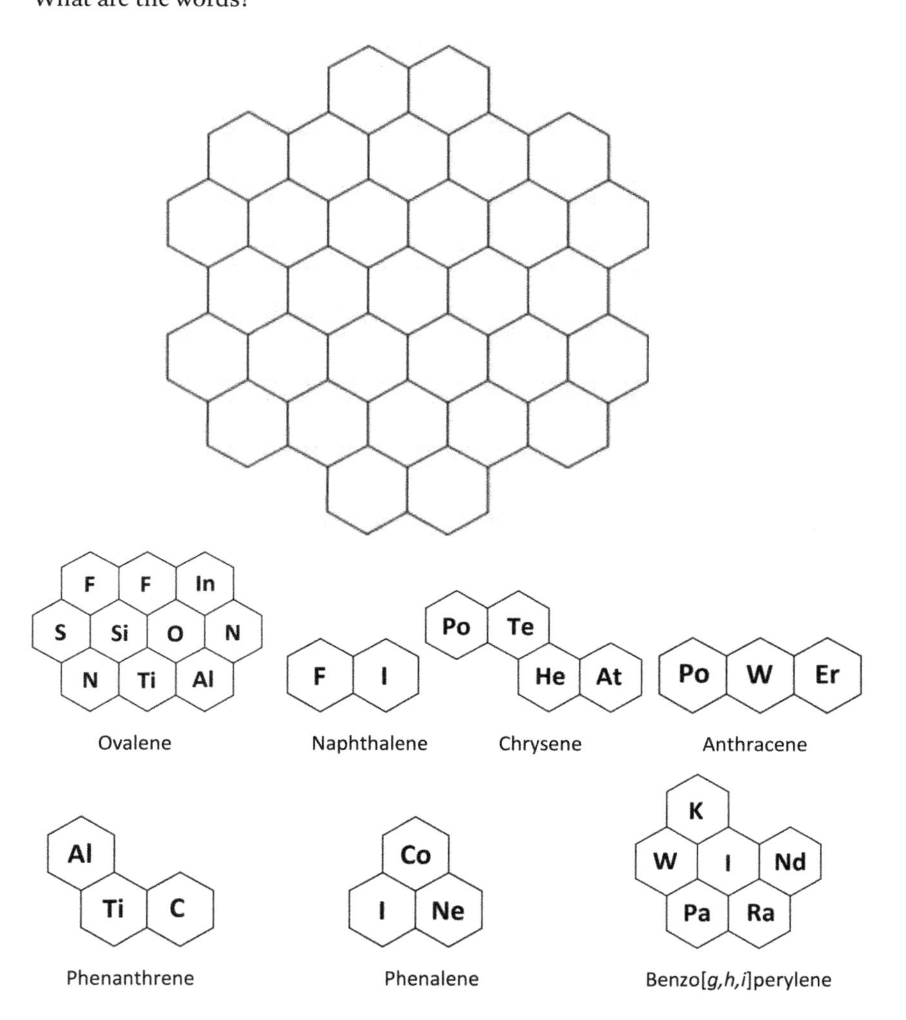

29. Finally, insert eight shapes (again based on PAHs) into the honeycomb grid above them so that the elemental symbols in each hexagon reading left to right spell out seven things you might find in a mine. No need to rotate any of the shapes. What are the words?

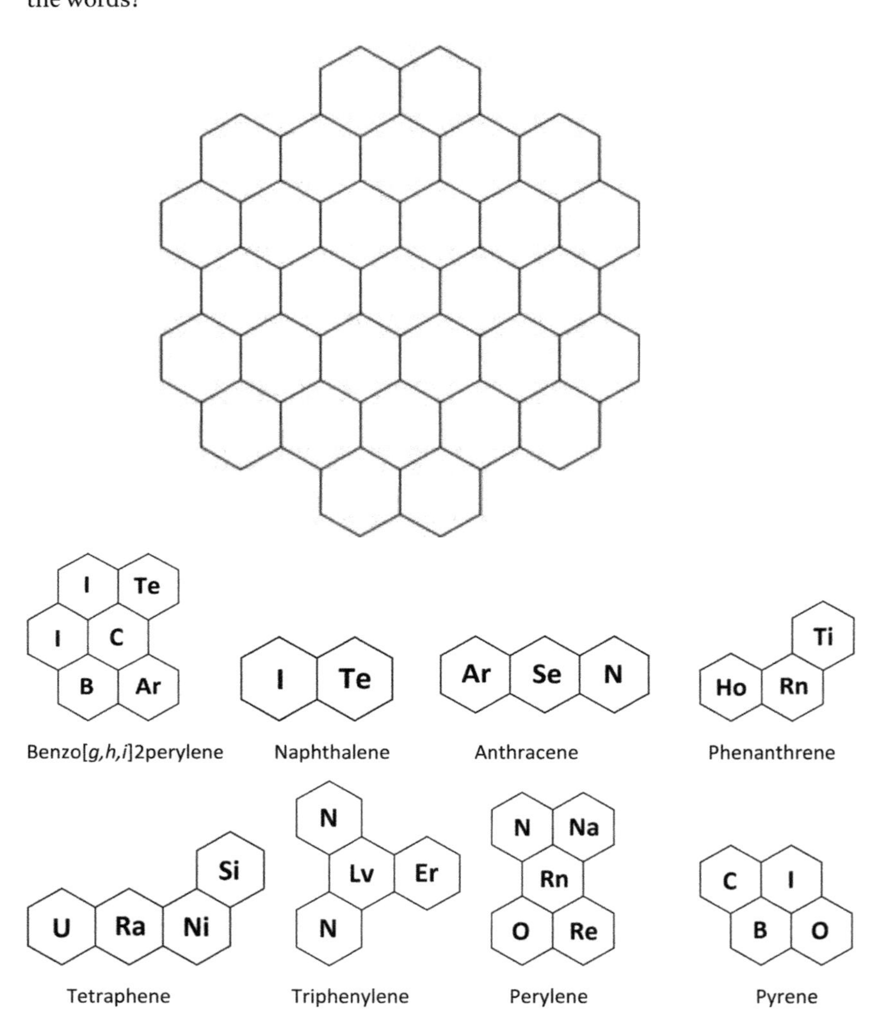

James Watson and Francis Crick with their model of DNA. A. Barrington Brown, © Gonville Caius College/ coloured by Science Photo Library

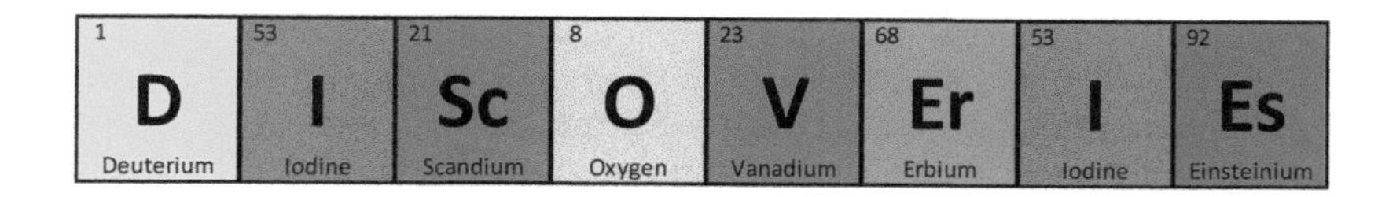

The real voyage of discovery consists, not in seeking new landscapes, but in having new eyes.

Marcel Proust (1871–1922), In Search of Lost Time

Clearly the above was written before laser eye surgery. The great inventor Thomas Alva Edison (1847–1931) is famously misquoted as saying '*Genius is one per cent inspiration and 99 per cent perspiration*' but there is an element of truth in it. Especially when you look at the laborious isolation of radioactive elements from tons of pitchblende by the Curies. Yet some discoveries are a matter of serendipity (such as Becquerel's original discovery of radioactivity in 1896), the word serendipity first being coined by the British politician Horace Walpole in 1754, referring to a Persian Fairy Tale *The Three Princes of Serendip*, the three princes '*always making discoveries, by accidents and sagacity, of things which they were not in quest of.*' (*Serendip* was an old name for Sri Lanka).

1. *Nature can be a sardonic jester at times, when you come to think of the hundreds of thousands of alchemists in the past few thousand years toiling and broiling over their furnaces, spending laborious days and sleepless nights trying to transmute one element into another, a base into a noble metal, and dying unrewarded in the quest, whilst we at McGill, by my first experiment, were privileged to see, in thorium, the process of transmutation going on spontaneously, irresistibly, incessantly, unalterably! There's nothing you can do about it. Man cannot influence in this respect the atomic forces of Nature.*

So wrote Frederick Soddy (1877–1956) about his discovery of a new element, produced from the radioactive decay of thorium. What was the element?

A. Radium
B. Radon
C. Rutherfordium

2. Acrostic. Solve the clues across in the left-hand grid and reading down in the first column, you will find the name of a chemist who was a prolific discoverer of elements (two of which are in the across clues). Transfer the letters in the left-hand grid into the right-hand grid, as specified, and you will find a quotation appropriate to this section and the name of the person who said it.

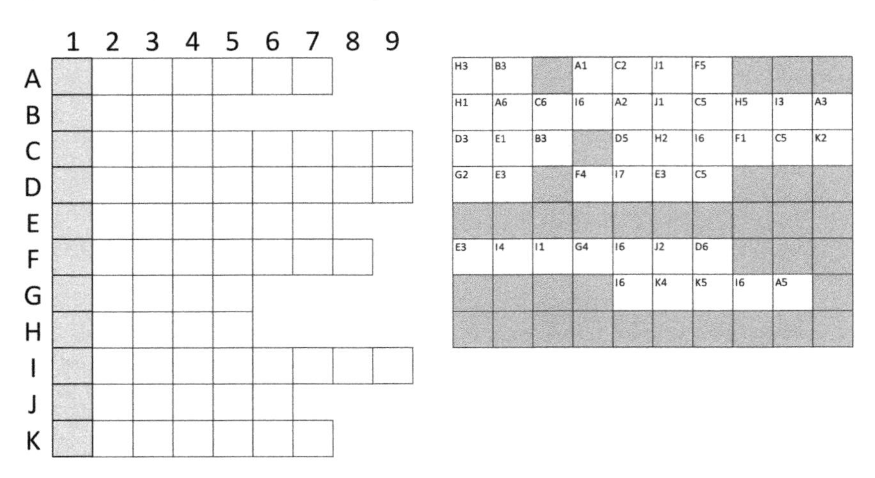

A. X-Ray crystallographer who won the 1964 Nobel Prize in Chemistry for her elucidation of the structures of biomolecules
B. US chemist who discovered deuterium for which he was awarded the 1934 Nobel Prize in Chemistry
C. One of the elements discovered by the chemist in the down clue
D. Another of the elements discovered by the chemist in the down clue
E. Transition metal element discovered in Copenhagen in 1923 and named after the Latin name for that city
F. Discoverer of argon, for which he won the 1904 Nobel Prize in Physics
G. British polymath who was instrumental in deciphering the Egyptian hieroglyphs on the Rosetta Stone and who has a modulus of elasticity named after him
H. British chemist and physicist who invented the vacuum flask
I. One of the two elements whose discovery was announced by Glenn T Seaborg on a *Quiz Kids* TV programme on 11 November 1945
J. Medication whose use in treatment of erectile dysfunction was accidentally discovered by Pfizer scientists originally looking for an angina drug in the 1990s
K. One of the four elements discovered in a Swedish village and named after it

3. Which transition metal element was discovered in 1802 by Swedish analytical chemist Anders Ekberg (1767–1813) who explained the reason for his naming of it thus:

To the new member among the metals I give the name _ _ _ _ _ _ _ _, partly to follow the custom of adopting names from mythology,† *and partly to allude to the fact that the oxide of this metal is incapable of feeding itself even in the middle of a surplus of acid.*

A. Tantalum
B. Titanium
C. Tungsten

4. Which British chemist won the 1904 *Nobel* Prize in Chemistry for his discovery of *noble* gases?

A. Lord Kelvin (William Thomson)
B. Lord Rayleigh (John William Strutt)
C. Sir William Ramsay

5. American inventor and industrialist Thomas Alva Edison (1847–1931) writes about the steps towards the invention of something that transformed the modern world:

All night, Batchelor, my assistant, worked beside me. The next day and the next night again, and at the end of that time we had produced one carbon out of an entire spool of Clark's thread. Having made it, it was necessary to take it to the glassblower's house. With the utmost precaution Batchelor took up the precious carbon, and I marched after him, as if guarding a mighty treasure. To our consternation, just as we reached the glassblower's bench the wretched carbon broke. We turned back to the main laboratory and set to work again. It was late in the afternoon before we had produced another carbon, which was again broken by a jeweler's screwdriver falling against it. But we turned back again, and before night the carbon was completed...

What was invented?

6. Which element was first synthesised by an American and Russian team in Dubna, Russia in August 2003?

A. Dubnium
B. Moscovium
C. Ruthenium

---

† The character the element was named after was a mythical king of Phrygia, son of Zeus, and was condemned for revealing secrets of the gods, to stand in water up to his chin, with fruit hanging above him. The water receded when he tried to drink and the fruit evaded his grasp.

7. Which non-stick polymer was accidentally discovered in 1938 by DuPont chemist Roy Plunkett when he noticed a white powder formed in cylinders storing the refrigerant gas tetrafluoroethylene?

8. Which rat poison's therapeutic use as a blood anticoagulant was investigated after an attempted suicide, and has also been implicated in the death of Joseph Stalin?

9. In 1875, the French chemist Paul Emile Lecoq de Boisbaudrun (1838–1912) discovered an element by spectroscopy and named it after his country, but was accused of naming it after himself, albeit in a roundabout fashion. What was the element?

10. Nitrocellulose was reportedly discovered by accident when the German-Swiss chemist Christian Friedrich Schönbein wiped an acid spillage (a mixture of nitric and sulfuric acid, HNO_3 and H_2SO_4) on his kitchen table with a cotton apron. Hanging the apron on the oven door to dry, it flashed and so began the explosive history of guncotton. Yet nitrocellulose was also to play a starring role as a base for photographic film. What was its name?

    A. Cellulite
    B. Celluloid
    C. Pyrocellulose

11. Match these discoveries with their year of discovery:

1. Electron	**A.** 1669
2. Potassium, Sodium	**B.** 1807
3. Higgs Boson	**C.** 1897
4. Tennessine	**D.** 1931
5. Deuterium	**E.** 2009
6. Phosphorus	**F.** 2012

12. Which substance was discovered with the assistance of sticky tape and led to the 2010 Nobel Prize in Physics being awarded to its discoverers Andre Geim and Kostya Novoselov?

13. Which invention was discovered by chance by Louis Daguerre (1787–1851) in 1835 due to a broken mercury thermometer in a cupboard?

14. William Hyde Wollaston (1766–1828) advertised his newly discovered element as below, in the window of a Soho shop in April 1803. What was the element?

_ _ _ _ _ _ _ _ _;

OR,

NEW SILVER,

HAS these Properties amongst others that shew it to be

A NEW NOBLE METAL.

1. IT dissolves in pure Spirit of Nitre, and makes a dark red solution.
2. Green Vitriol throws it down in the state of a regulus from this solution, as it always does Gold from *Aqua Regia.*
3. IF you evaporate the solution you get a red calx that dissolves in Spirit of Salt or other acids.
4. IT is thrown down by quicksilver and by all the metals but Gold, Platina, and Silver.
5. ITS Specific Gravity by hammering was only 11.3, but by flatting as much as 11.8.
6. IN a common fire the face of it tarnishes a little and turns blue, but comes bright again, like other noble metals on being stronger heated.
7. THE greatest heat of a blacksmith's fire would hardly melt it;
8. BUT if you touch it while hot with a small bit of Sulphur it runs as easily as Zinc.

IT IS SOLD ONLY BY

MR. FORSTER, at No. 26, GERRARD STREET, SOHO, LONDON.

In Samples of Five Shillings, Half a Guinea, & One Guinea each.

A Geology Lesson from the Jurassic Coast (photo by the author).

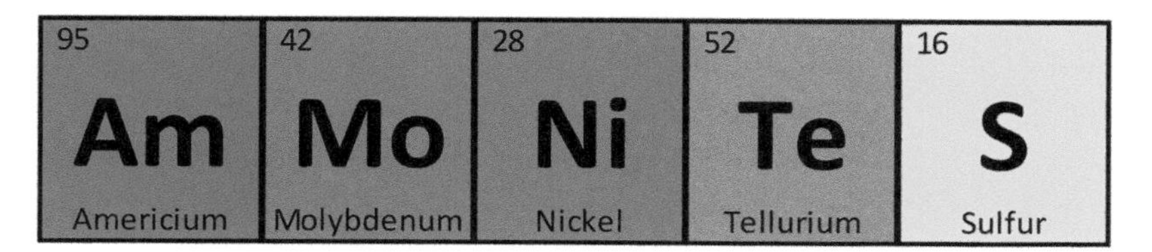

Earth Sciences

As a young man, like many, the author was fascinated by the ammonites he found on occasional seaside jaunts to Charmouth on Dorset's Jurassic Coast (and took his own children looking for ammonites on Kilve beach on the North Coast of Somerset). After studying physical geography at school, he was re-introduced to the wacky world of the Earth Sciences later in life due to a mid-life career change, and in the process acquired a pair of ammonite cufflinks, a pair of trilobite cufflinks and several other pairs of various coloured and variegated stone ones (like Welsh slate and green malachite), occasionally daring to wear an ammonite on one sleeve and a trilobite on the other, representing the Jurassic and Cambrian Periods respectively and spanning hundreds of millions of years between them.

1.

Ta	Ce	O
Re	S	Te
La	C	U

Using all 9 elemental symbols in the grid above (and *not* separating the letters of a two-letter symbol), spell out a Geological Epoch (two words).

Then answer these questions using some of the symbols always including the middle symbol and not switching letters of 2-letter symbols (using each symbol once and once only in each word):

A. Fine-grained metamorphic rock that cleaves easily and is mined in Wales (5)
B. Cylindrical rock samples (5)
C. First asteroid discovered, on 1 January 1801, by Giuseppe Piazzi which lends its name to an element discovered 2 years later (whose symbol is in the grid above) (5) What is the element? (6)
D. Metallic minerals such as bauxite (4)
E. _ _ _ _ _ of Matter (such as gas, liquid, solid, plasma) (5)

2. The following Earth Sciences words begin with and end with the same two letters (in the same order) in each word (but not the same two letters throughout). To make it easier, the words are listed in alphabetical order. Complete the words:

A. _ _ LUVI _ _ — Describing sediment that has been deposited by a river

B. _ _ ISI _ _ — Geologic age

C. _ _ DROGRAP _ _ — Science of water bodies

D. _ _ G _ _ — Molten or semi-molten material that forms igneous rocks

E. _ _ N _ _ — Dinosaur

F. _ _ OGE _ _ — Geologic period

G. _ _ NNANTI _ _ — Mineral named after the English chemist who discovered iridium and osmium

3. Which element is normally rare in the Earth's crust but has anomalously high concentrations in rock strata dating around 66 million years ago (also known at the K–T boundary after the Cretaceous and Tertiary periods, the K referring to the German), pointing to a possible extra-terrestrial impact event which could have led to the dinosaur extinctions?

4. An elemental Wordsearch. Look for the following words from the Earth Sciences, up, down, diagonally or reverse:

P	Mo	No	Cl	I	N	I	C	H	I	Na	C	La	Y	Ds
Cu	Ac	Sc	As	P	Ba	S	Al	Ti	C	La	V	As	Nb	Ra
Te	B	H	S	Li	Ge	Ga	Ne	Ta	Hg	I	Lv	Ar	I	Ca
Al	N	I	Y	O	In	O	Si	Ni	Bk	Es	Am	K	Rb	B
F	O	Na	C	Ce	Ge	Ar	Lv	Te	Np	K	Zn	O	Co	Ra
Fm	Md	Ir	N	Ne	P	B	O	S	S	Er	Ni	Se	Pr	C
Ac	Ge	Ds	Y	Ti	C	H	Al	K	Y	F	B	Zr	O	H
Cs	O	Pb	Y	Eu	Te	Lr	Al	U	Er	Po	O	H	Li	I
Ar	P	Zn	U	Ca	Li	Te	Tl	O	C	Ts	Y	S	Te	O
B	H	Ce	Rg	U	S	Ca	U	Se	S	Ar	Ru	Sn	S	Po
Na	Y	Am	Mo	Ni	Te	S	Dy	Na	Mo	Au	Ho	Ho	Fl	Ds
N	Si	Zr	B	Dy	In	Th	Ra	Md	S	Zn	Ru	Sb	Hg	Ag
I	Cs	Ti	Ra	Er	Rg	U	Es	I	Tm	As	Yb	S	Nd	Re
C	Re	Ta	Ce	O	U	S	Lv	Cl	I	No	C	La	Se	Es
Sc	Y	Ra	Hs	Cn	Si	Si	C	Al	Ca	Re	O	U	S	O

- Amber
- Ammonites
- Arkose
- Basaltic Lavas
- Brachiopods
- Calcareous
- Carboniferous
- Chalky
- China Clay
- Cinnabar
- Clinoclase
- Coprolites
- Cretaceous
- Cubic
- Erratics
- Esker
- Geophysics
- Monoclinic
- Pliocene
- Pachycephalosaurus
- Silvisaurus
- Tennantite
- Titanite
- Znucalite

5. Which dark green arsenate mineral was only discovered in 2020 in a specimen at the Natural History Museum (but originally from Wheal Gorland Mine in St Day, Cornwall which is now built over) and named after the Cornish language name for the county?

   A. Kaolin
   B. Kernowite
   C. Kryptonite

6. Which common compound was named after a Roman god (one theory being that the salt of this compound was extracted from camel dung and the remains of sacrifices outside his shrine)? This god also gives its name to a commonly found fossil. What is the fossil?

7. Link these metals to their ores:

1. Aluminium	**A.** Bauxite
2. Copper	**B.** Cinnabar
3. Iron	**C.** Galena
4. Lead	**D.** Haematite
5. Mercury	**E.** Ilmenite
6. Titanium	**F.** Malachite

8. Replace the 3 question marks with letters to form a mineral, from the four letters in the middle intersections, so that the letters in each of the 3 full circles also spell out minerals (clue – they all contain calcium):

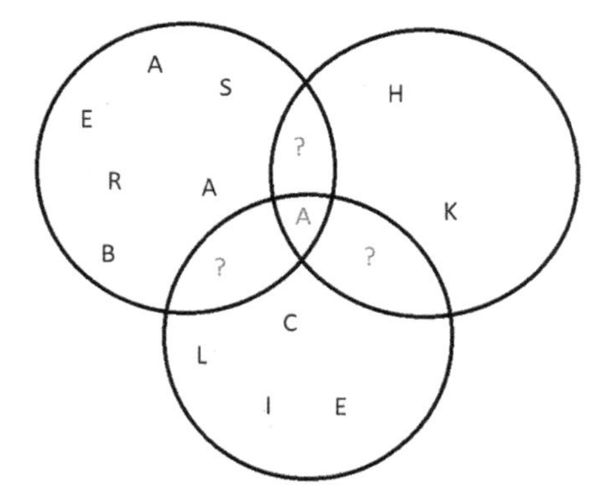

9. What yellow or smoky mineral, a form of zircon, is a word that means professional mumbo jumbo?

The author at Giant's Causeway

10. Giant Steps

Northern Ireland's famous Giant's Causeway is a massive formation of polygonal basaltic columns. These were created by natural processes about 50–60 million years ago. How were the polygons formed?

- **A.** By a macroscopic manifestation of the hexagonal-close packing (HCP) arrangement of the atoms in the crystals of the rock (like you see in peas in a boiling saucepan)
- **B.** Contraction of cooling lava causing fracturing (like you see in drying mud)
- **C.** They were built by an Irish giant who was challenged to fight a Scottish giant and he built the causeway so they could meet
- **D.** None of the above

11. Imagine the images below are some of the basaltic polygonal structures (end-on). Using the elemental symbols (and not switching the 2 letters in each symbol), place each of the seven elements in the hexagons to spell out three 6-letter words reading diagonally. In each of the four puzzles there will be a shared element in the middle hexagon which when combined, and reading across, will spell out a geological feature that may be seen at the Giant's Causeway.

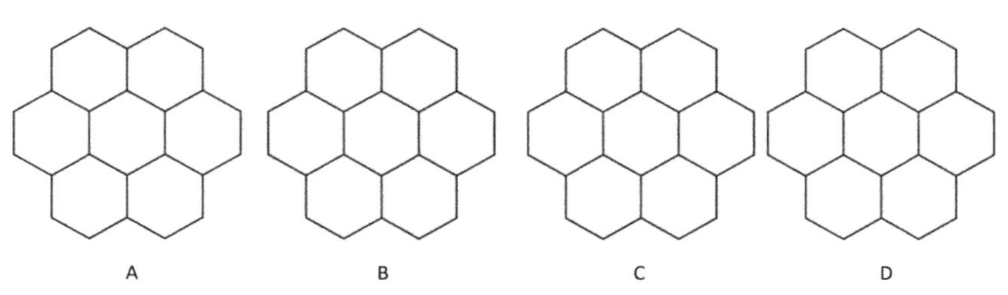

Elements:

A. Al, Ce, Co, Er, Ra, Si, Ts
B. Be, Ca, Cu, Fe, Ra, Re, Te
C. At, Ce, Cs, Na, No, Pa, Ti
D. C, He, Lu, Ni, Re, Se, Ts

12. As mineral-rich running water flows into caves, it can form deposits which build up to form structures called stalactites and stalagmites. What is the mineral that forms these structures?

13. The Earth's core is composed primarily of which metallic element?

14. Which element, whose name comes for the Greek for heavy, is used to add extra weight in compound form to drilling muds for oil wells?

15. Thanks to the British geologist Arthur Holmes (1890–1965) and uranium–lead radiometric dating, we now have an estimate of the age of the Earth. What is it?

A. 2.5 billion years
B. 3.5 billion years
C. 4.5 billion years

1884 Edition of the 'Puck' magazine (published in New York). © Library of Congress/Science Photo Library

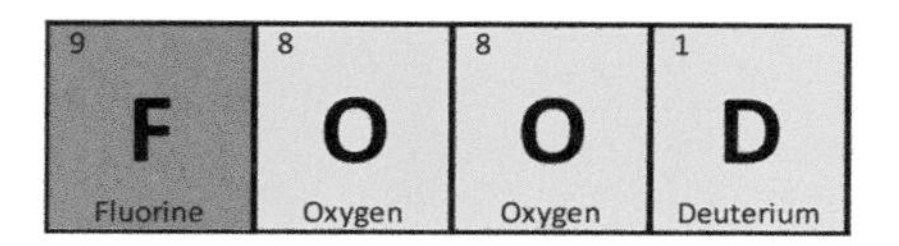

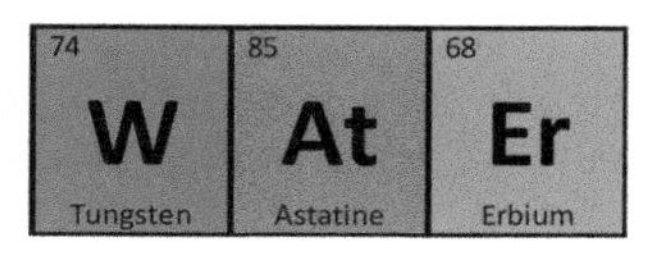

The author was a food analyst for several years of his career 'at the bench' and has many fond memories using what was then seen as the Bible of the food analyst '*Pearson's Composition and Analysis of Foods*' and the nutritionist's Bible '*McCance and Widdowson's The Composition of Foods.*'

1. Which compound used to treat malaria may be found in tonic water?

   A. Ascorbic acid
   B. Citric acid
   C. Quinine

2. Which type of cheese is also a type of subatomic particle?

3. Which alkaline earth element connects chalk and cheese?

4. Which popular soft drink was originally named after an element and advertised with the slogan '*Made in Scotland from Girders?*'

5. $CH_2{=}CH_2$ is a gaseous compound that is naturally involved in the ripening of fruit (the = signifying a double bond) but it is also synthesised in commercial fruit-ripening rooms to speed up the process. What is the compound's name?

6. Which compound causes the bad smell of rotten eggs?

7. Drinking water can have more or fewer dissolved minerals (usually calcium or magnesium salts) in it, dependent upon the hydrogeology of the rocks it permeates through. What are the two common words used to describe water that has more or fewer dissolved minerals, respectively?

8. Replace the 3 question marks with letters to form a food condiment from the four letters in the middle intersections, and so that the letters in each of the 3 full circles spell out (UK) permitted food sweeteners.

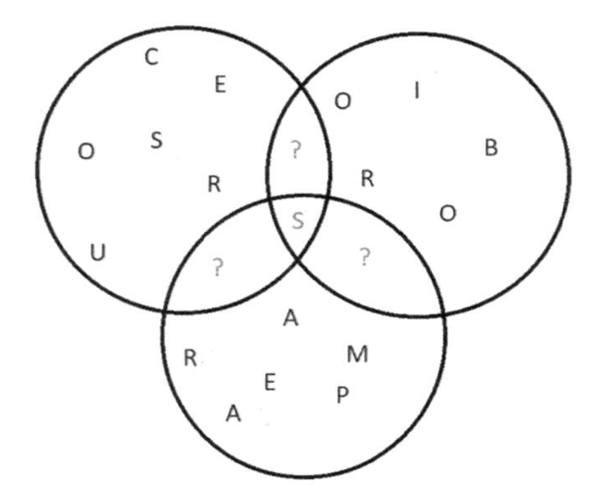

9. Fill in the name of the missing compound in this chemical equation:

$$\underset{\text{Ethanol}}{CH_3CH_2OH(l)} + \underset{\text{Oxygen}}{O_2(g)} \rightarrow \underset{?}{CH_3COOH(aq)} + \underset{\text{Water}}{H_2O(l)}$$

10. Before the advent of dietary supplements, the only major source of vitamins was foods. The word 'Vitamin' comes from a concatenation of the words 'Vital' and 'Amine' in 1912 by the Polish biochemist Dr Kasimir Funk (1884–1967), 'vitamine' now known to be a misnomer and the 'e' dropped, although of course they are vital to life.

Vitamins are usually referred to by letters (and some also with numbers), but they also have other names. Which of these compounds is *not* a vitamin:

A. Ascorbic acid
B. Benzoic acid
C. Biotin
D. Nicotinic acid
E. Riboflavin
F. Thiamine

11. Which element is implicated in the tear-jerking effects of onions?

A. Iron
B. Phosphorus
C. Sulfur

12. Chemist Alfred Bird (1811–1878) was a Fellow of the Chemical Society and an inventor of a water barometer. Yet he is better known for another invention that he prepared for his wife whose digestive system could not tolerate eggs or yeast. What was the invention, which you can still buy today?

13. In the 18th century *Devon Colic* was a sometimes fatal condition in the English county that was traced to lead poisoning in a common local beverage. What was the beverage?

A. Beer
B. Cider
C. Tea

14. Which toxic compound smells of bitter almonds and is released when the kernels of apricots, peaches and cherries are digested?

15. During the First World War, the American public were encouraged to save discarded peach stones and certain other food waste as part of the war effort:
Girl Scouts and Boy Scouts got involved:

'*Gather up the peach pits,*

Olive pits as well.

Every prune and date seed,

Every walnut shell.'

What was the purpose (and it's nothing to do with the toxic compound in the previous question)?

16. Unscramble these words to reveal permitted Food Additives† with their functions given by way of assistance:

A. ACE TENOR	Colouring
B. A FAKE MUSCLE (2 words: 10, 1)	Sweetener
C. ANDROID ICEBOX (2 words: 6, 7)	Propellant and acidity regulator
D. A ZINC BODICE (2 words: 7, 4)	Preservative
E. INCOME RAIDING (2 words: 6,7)	Colouring
F. LOUSE CELL	Thickening agent
G. NEAR CARNAGE	Thickener and gelling agent
H. NICE STYLE (2 words: 1-8)	Reducing agent in baking
I. SAME APART	Sweetener

† As at date of publication, permitted in the UK.

17. Which well-known palindromic UK brand originated from the Liebig's Extract of Meat Company (originally named after the German organic chemist Baron Justus von Liebig (1803–1873))?

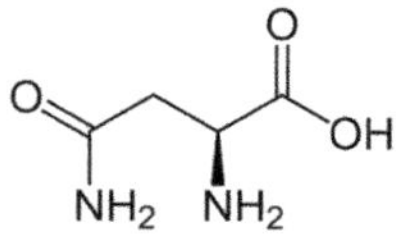

18. Above is the structure of the first amino acid to be isolated (in 1806) from a vegetable which can make your urine smell after eating. What is the vegetable?

19. Which commonly eaten vegetable contains small amounts of the toxic alkaloid *solanine* (also found in Deadly Nightshade) which is more prevalent when the vegetable is green or green in parts (and should not then be eaten)?

20. Which well-known carbonated soft drink began its life in 1920 and was originally called 'Bib-Label Lithiated Lemon-Lime Soda,' the word 'Lithiated' referring to one of the ingredients – lithium citrate?

21. Fizzy drinks were made possible by the invention of carbonated water in 1767 by an English chemist. What was his name?

A. Henry Cavendish
B. Humphry Davy
C. Joseph Priestley

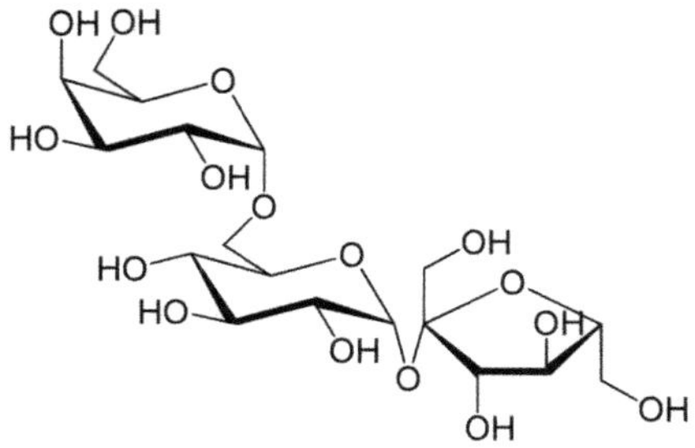

22. Carbohydrates are so named since generally their formulae are a multiple of one carbon atom to one water molecule. The above carbohydrate is present in beans and some other foods. Since it is not digested by the acids in the stomach, it is broken down by bacteria in the intestines to produce below-belt breezes. What is the compound's name?

A. Fructose
B. Raffinose
C. Sucrose

23. John Stith Pemberton (1831–1888) was an American pharmacist and veteran. Having sustained a sabre wound in battle, he experimented with painkillers and other compounds to cope with his addiction to morphine and came up with the recipe for a very famous soft drinks brand. What is it?

24. The Japanese delicacy known as *fugu* is potentially lethal if not prepared properly due to the presence of a potent neurotoxin known as tetrodotoxin. It is found in the liver, skin and gonads of which fish?

25. What compound, one of several responsible for the heat of chillies is the active ingredient of pepper spray?

A. Capsaicin
B. Piperine
C. Dihydrocapsaicin

26. Some compounds with the same formulae and structure differ only in that they are exact mirror images of each other. These are known as optical isomers (or enantiomers), since they rotate the plane of polarised light either one way or the other. Such an example is carvone which is the flavour component in some herbs, either as *d*-Carvone or *l*-Carvone (for dextro- or laevo-rotatory respectively). Each enantiomer imparts a different taste. What two herbs may they be found in?

A. Parsley and sage
B. Dill and spearmint
C. Rosemary and thyme

27. L-Cysteine is an amino acid and (UK) permitted food additive (used in baking) which may be sourced from what?

A. Human hair
B. Pig hoof
C. Poultry feathers

28. The American pharmacist Wilbur Scoville (1865–1942) wrote *The Art of Compounding* which ran into eight editions. However, he is perhaps more famous for his Scoville Scale. What is it a measure of?

29. Which vegetable, when boiled, produces a coloured water which may be used as a pH indicator?

30. Bergamottin is an example of a furanocoumarin which, by interfering with one of our body's enzymes, can affect the breakdown of certain prescribed drugs. It is present in a certain type of fruit that should therefore not be eaten when these drugs are taken. What fruit?

A. Grape
B. Grapefruit
C. Gooseberry

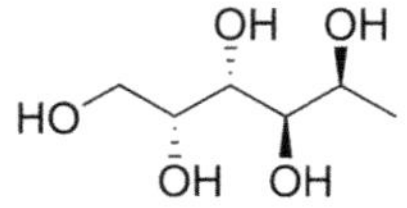

31. The IUPAC name for the above compound is (2*R*,3*S*,4*R*,5*S*)-hexane-1,2,3,4,5-pentol has a rather more common name of fucitol. What spice may it be found in (as the enantiomer L-fucitol)?

A. Chilli
B. Ginger
C. Nutmeg

32. There are 21 amino acids common to all life forms, and for humans there are 9 that are classed as essential (*i.e.* they must come from the diet). These 9 are *histidine*, *isoleucine*, *leucine*, *lysine*, *methionine*, *phenylalanine*, *threonine*, *tryptophan* and *valine.* Which food cultivated by the Incas and still popular today contains all 9 amino acids in addition to omega fats, B vitamins and minerals?

33. Deli Bal or 'Mad Honey' is a honey found in the high Turkish mountains by the Black Sea, with the psychotropic ingredients grayanotoxins (rhodotoxin, andromedotoxin and acetylandromedol) from certain species of which plant?

A. Geranium
B. Fuchsia
C. Rhododendron

34. In the 18th century, the potency of British Royal Navy rum was verified by mixing the rum with a pinch of gunpowder and lighting a match. If the gunpowder still ignited, the rum had sufficient alcohol (ethanol) to pass muster. This test has given rise to a word that still appears on liquor bottles today signifying the alcohol content. What is the word?

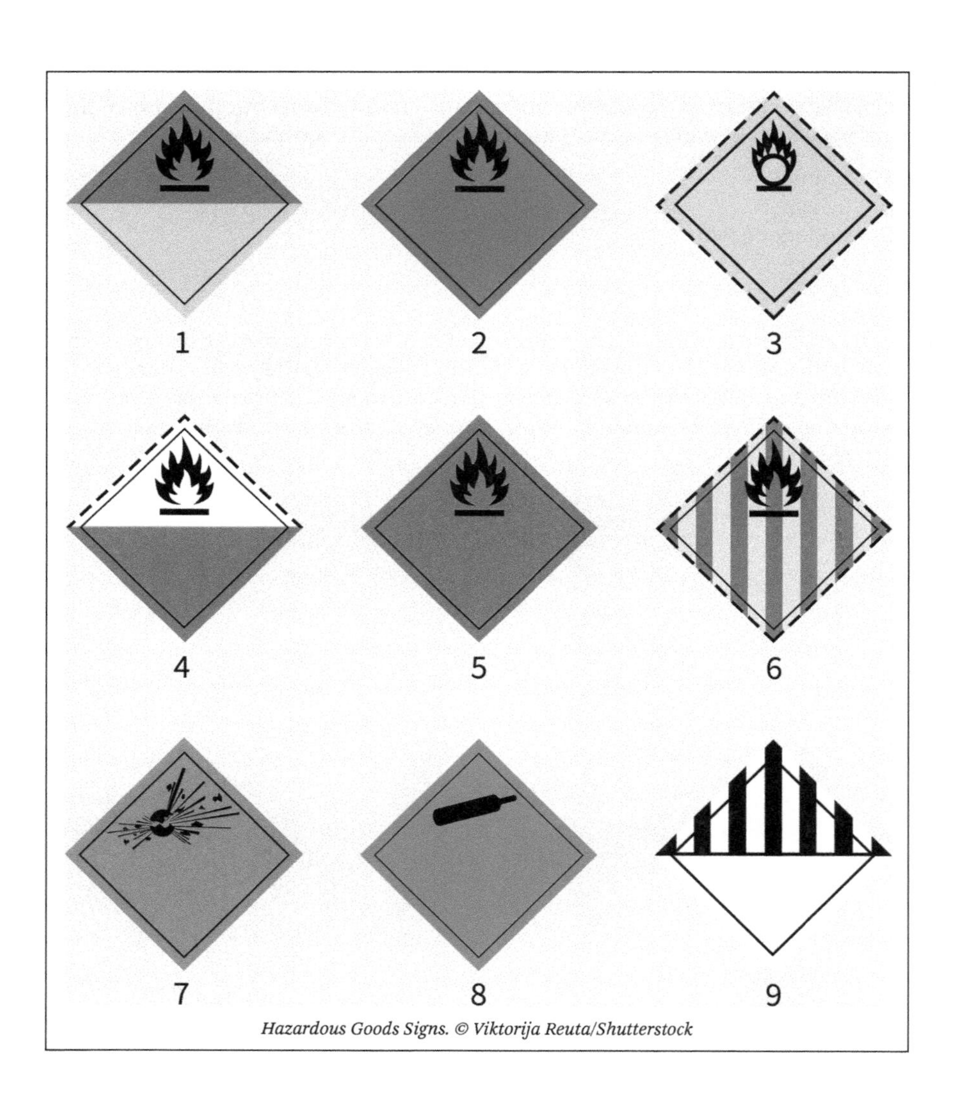

Hazardous Goods Signs. © Viktorija Reuta/Shutterstock

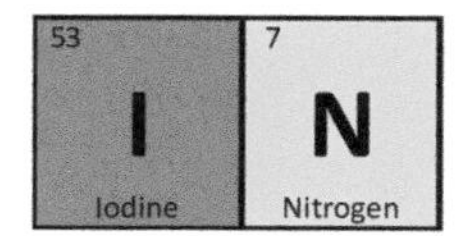

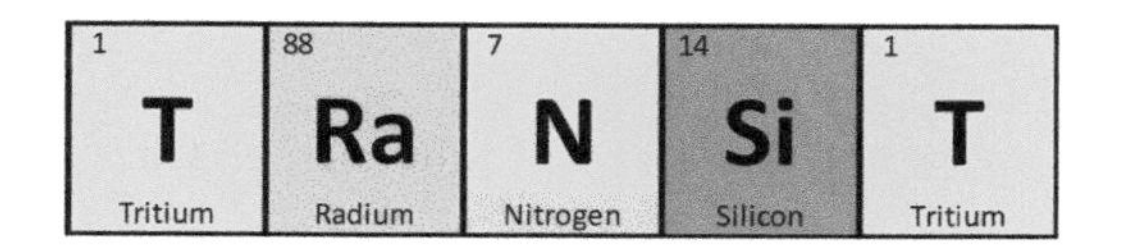

Getting from A to B? Perhaps nothing could be as difficult to predict, as the relentless advance of transportation over the last 250 years. From horse drawn carriages (and the futurologists' concern of cities being buried in horse manure as the populations grew), canals, railways, the internal combustion engine, and more recently airliners flying on bio-fuel, commercialised space travel, electric cars, and autonomous vehicles. What will the next 100 years bring? Beam me up, Scotty?

1. Link the nine Hazardous Goods Signs opposite to their meanings:

   A. Dangerous when wet
   B. Explosive
   C. Flammable gas or Flammable liquid
   D. Flammable solid
   E. Miscellaneous dangerous goods
   F. Non-flammable non-toxic gas
   G. Organic peroxide
   H. Oxidising agent or oxidising gas
   I. Spontaneously combustible

2. Which element was Henry Ford (1863–1947) talking about here? Regarding a special type of lightweight steel alloyed with this element for the production of the Model-T motor car:

   But for _ _ _ _ _ _ _ _ there would be no automobiles!

   A. Cobalt
   B. Manganese
   C. Vanadium

3. Which element did US chemist and inventor Charles Goodyear (1800–1860) use to vulcanise natural rubber, crosslinking the carbon chains to make a durable rubber suitable for tyres?

   A. Iron
   B. Phosphorus
   C. Sulfur

4. Lead was once used extensively in petrol in the form of tetraethyl lead which was an antiknock additive. This has now been phased out but lead is still used in many road-going vehicles under the bonnet. Where?

5. Lithium iron phosphate, $LiFePO_4$ is an essential compound for some modern vehicles. Where is it used?

6. Sodium azide, NaN_3 is an explosive. Yet it may be found in safety features of most modern road vehicles. What is the safety feature?

7. Which metallic element, which burns with a very bright light, has been used in flash bulbs and incendiary bombs but has also been used to build top of the range racing bikes?

8. Where might you find a precious metal in your road vehicle (fossil fuel engine) and what is the metal?

9. Trichloroethylene (IUPAC name trichloroethene) was once used as a degreaser and to clean rocket engines. One of its common names is also a means of transport. What is it?

10. The krypton isotope Kr-85 has been monitored in the atmosphere to detect nuclear fuel reprocessing activities, but what is KR175?

11. On December 1, 1783, just ten days after the first manned free-hot air balloon flight (made by the Montgolfier Brothers), the first manned gas balloon flight took place. What gas was used?

    A. Helium
    B. Hydrogen
    C. Methane

12. Chromium compounds produce a wide variety of paint pigments, one of which is Chrome Yellow. This colour is well known to previous generations of American school children? Why?

13. The German organic chemist Friedrich August Kekulé (1829–1896) reportedly had a dream on a Clapham omnibus that helped him elucidate the ring structure of a building block of organic chemistry. What was the molecule?

14. What could you call this compound?

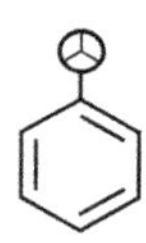

15. Despite their name, the early German submarines known as U-Boats were not powered by uranium (or any other fissile material). The anglicised name is from the German U-Boot, short for *Unterseeboot* (undersea boat).

Nuclear-powered submarines now use uranium-235. The first nuclear-powered submarine was *USS Nautilus* (named after Captain Nemo's submarine in Jules Verne's sci-fi novel *Twenty Thousand Leagues Under the Sea*). *USS Nautilus* was the first submarine to complete a submerged transit of the North Pole (this transit being dubbed *Operation Sunshine*). What year did it do this?

A. 1950
B. 1955
C. 1958

16. What metal were the first lightweight 'alloy wheels' made out of?

17. The new car smell comes from substances known as VOCs (or Volatile Organic Compounds) such as toluene, ethyl benzene, xylenes, trimethylbenzenes and some alkanes. They come from a variety of sources including plastics, upholstery, adhesives and lubricants. What potential smell may be found in some electric cars?

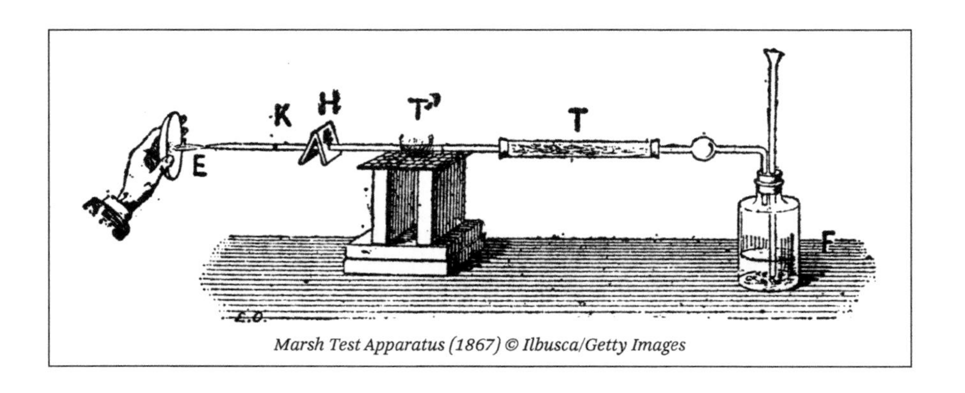

Marsh Test Apparatus (1867) © Ilbusca/Getty Images

The term *Litmus Test* has made it into modern parlance. The original Litmus Test was for the testing of alkalinity and acidity, using coloured extracts of lichens, and the word litmus comes from the Middle Dutch from *lak* meaning a dark red resin and *moes* meaning pulp.

Litmus paper still employs the chromophores extracted from lichens and will change colour dependent on the alkalinity or acidity of the solution it is immersed in. Litmus was first used by Arnaldus de Villa Nova (c1240–1311), an Aragonese physician, alchemist and astrologer.

Important though acidity and alkalinity are, chemical analysis now has a whole arsenal of techniques at its disposal, from traditional 'wet chemical' analysis (such as distillation, paper chromatography, spot tests and titration) to quite sophisticated and sometimes hyphenated methods, such as GC-MS, coupling Gas Chromatography with Mass Spectrometry.

1. In 1909 Danish chemist Søren Sørensen introduced a (now common) analytical chemistry concept while he was head of the prestigious Carlsberg Laboratory in Copenhagen. What was it?

2. Name the alkali and alkaline-earth metal elements which produce the following colours in flame tests:

   A. Intense yellow (alkali metal)
   B. Lilac (alkali metal)
   C. Red-violet (alkali metal)
   D. Red (alkaline-earth metal)

3. The words *accurate* and *precise* are synonyms in everyday language, but in science and particularly chemical analysis they mean quite different things. *Accurate* means closeness to true value. *Precise* means closeness to each other. So a set of precise measurements may still be inaccurate, off true, or off target, like this illustration using a target as an analogy:

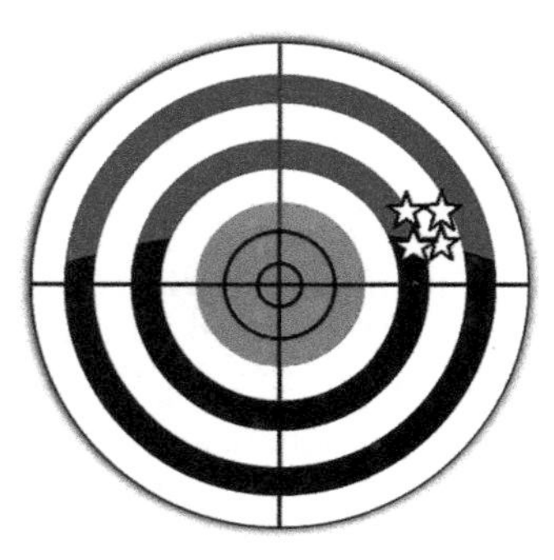

So what are these measurements:

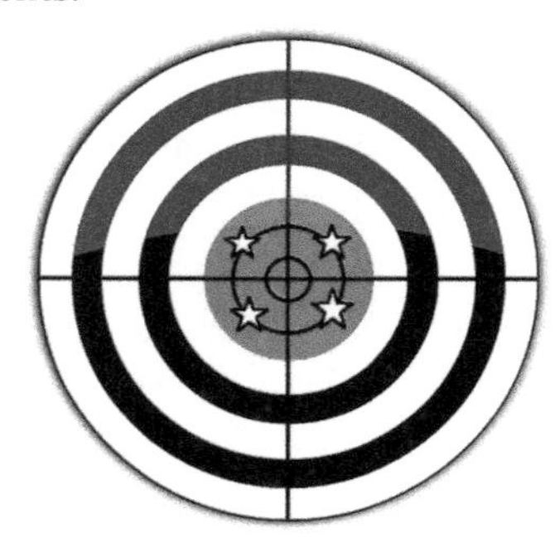

A. Accurate and precise?
B. Accurate and imprecise?
C. Inaccurate and imprecise?

4. Over 100 years ago Russian botanist Michel Tswett (1872–1919) developed the analytical technique of *chromatography,* initially to separate plant pigments (hence the word including *chroma-*, Greek for colour although the botanist's last name is also the Russian for colour). His gravestone has a Russian inscription which translates as '*He invented chromatography, separating molecules but uniting people*.'[†] Since Tswett's work (involving column chromatography), this technique has come on in leaps and bounds and is a key tool for analytical chemists, most of it now using instrumentation. The author has used many of these techniques, most of them shortened to acronyms. One such technique is HPLC. What does HPLC stand for?

   A. Hewlett-Packard Liquid Chromatography
   B. High Performance Liquid Chromatography
   C. Hydrogen Peroxide Liquid Chromatography
   D. Hydrolysed Protein Liquid Chromatography

5. The word 'normal' means different things to different people. Especially if you're a scientist (or Bill Bailey[‡]). Which of these descriptions would a scientist recognise as 'normal?'

   A. The concentration of a solution (Chemistry)
   B. A distribution of data around the mean (average) (Statistics)
   C. The component of a force perpendicular to the surface that an object contacts (Physics)
   D. Imaginary line forming a right angle with the tangent to a curved surface at a particular point (Microscopy)

6. The Marsh Test was the undoing of many a Victorian poisoner. Named not after the New Zealand crime writer Ngaio Marsh (who herself had a predilection for various poisons in her novels), this highly sensitive test was named after the chemist James Marsh (1789–1846) and was first published in 1836. It continued to be used in forensic toxicology (albeit with various modifications) until the 1970s. The test was described at some length in Dorothy L Sayers' crime novel *Strong Poison.* What toxic element was it originally designed to detect?

[†] The author recalls his first chromatography lesson in a science class at school, separating the plant pigments chlorophyll (green) and xanthophyll (yellow) on filter paper after using a solvent and a pestle and mortar to extract them from grass.

[‡] Bill Bailey is a British actor, comedian and musician. His 2022 UK Tour was entitled '*En route to normal.*' He is also a keen ornithologist.

7. In modern parlance, TLC means Tender Loving Care. It also stands for an analytical technique. What is it?

   A. Thin Layer Chromatography
   B. Tuned Laser Chemistry
   C. Tuned Light Chemistry

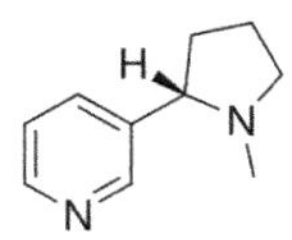

8. Belgian chemist Jean Stas (1813–1891) was the first person to develop a method for detecting vegetable alkaloids in body tissue, providing evidence in 1850 that the Belgian Count Hippolyte Visart de Bocarmé had killed his brother-in-law with the alkaloid above. The Count was executed by guillotine the following year. What is the alkaloid?

   A. Nicotine
   B. Morphine
   C. Strychnine

9. Reveal these pieces of laboratory equipment from their anagrams:

   A. Rye gins
   B. Tired hips (2 words)
   C. Value coat
   D. Comic prose
   E. Curing feet
   F. Goblin residence (2 words)

10. Similar to a word ladder, each answer below provides letters that must appear in the answer immediately following it. The puzzle's final answer is a piece of laboratory equipment.

A.	_ _ _	Expire (3)
B.	_ _ _ _	What you eat (4)
C.	_ _ _ _ _	Weary (5)
D.	_ _ _ _ _ _	IOU (6)
E.	_ _ _ _ _ _ _ _ _	Authorises (9)
F.	_ _ _ _ _ _ _ _ _ _	Lab equipment (10)

Co	K	La
Al	S	F
C	N	I

11. Using all 9 elemental symbols in the grid above (and *not* separating the letters of a two-letter symbol), spell out a common piece of laboratory glassware (two words) invented by German chemist Emil Erlenmeyer (1825–1909).

Then answer these questions using some of the symbols always including the middle symbol and not switching letters with 2-letter symbols (using each symbol once and once only in each word):

A. Fossil fuels (5)
B. Parts of a fish (4)
C. Body covering (4)
D. Sides (6)
E. Tricks (4)
F. Male poultry (5)
G. Loose (5)
H. Unwell (4)

12. Hanjie! A *nanogram* is 10^{-9} grams. A *nomogram* is a graphical calculating device. A *nonogram* however is a type of Japanese puzzle, named after one its inventors, *Non Ishida.* It is also known as *Hanjie* (from the Japanese for picture puzzle). The puzzle below is a variation on this.

Instead of creating a picture, it will reveal five things you might see in a laboratory, revealed reading across from the elemental symbols remaining after you have shaded out those squares as instructed by the sequence of numbers above the columns or beside the rows.

The numbers give the length of each consecutive run of shaded squares with a gap of at least one empty square between each run in the same row or column. For instance, if the numbers 1 3 2 appear next to a row, there must be 1 grey square followed by at least one white square, followed by 3 grey squares, followed by at least one white square and then followed by 2 grey squares. The fourth row has been completed by way of example.

	6	1 1 1 2	1 2 1	3 1 1	1 1 1	1 2 1	1 2 1	1 1 1	2 2 1
1 1 1	B	F	U	C	Ar	H	O	N	Er
1 2 1 2	Re	Fl	Og	S	As	Ru	K	Au	Zn
2 1 1 1	Bi	N	Li	Ir	Tm	U	Ce	S	C
1 3 2		Pa				P			Er
1 1 1 1	Ba	B	Y	U	N	Dy	Se	N	Es
2 1 1 1	F	O	B	Eu	U	P	Rn	Er	No
1 1 1 1	Ta	S	Te	P	Be	At	S	U	La
1	P	O	Re	C	I	S	I	O	N
1 1 1 2	Ba	Th	La	Rf	N	Pu	Ce	P	S

13. Which law, underpinning many analytical chemistry techniques, was discovered by French polymath Pierre Bouguer while looking at a glass of red wine while he was on holiday in Portugal?

A. Beer–Lambert Law
B. Bouguer–Beer Law
C. Bouguer's Principle

Mad Hatter statue, Llandudno, North Wales

Some work is more dangerous than others and several occupations exposed and expose workers to both physical and chemical hazards.

One of the early investigations into work-related diseases occurred in the 18th century. In organic chemistry, *aromatic* compounds are defined as compounds containing the benzene ring (C_6H_6). *Polycyclic aromatic hydrocarbons* (or PAHs for short) are compounds containing two or more benzene rings fused together. The carcinogenic effects of these compounds (though the compounds' identities were not known at the time) on chimney sweeps' young assistants (called 'climbing boys') was perhaps the first epidemiological study. *This prompted the author to compose the following risible rhyme:*

Ode to PAHs—or Sir Percival Pott's Postulations

Sir Percival Pott
Did quite a lot in 1775.
Way before The British law
Could keep a child alive.

A climbing boy
With "sooty warts"
(Actually, cancers, scrotal)
Begat Epidemiology
(Far more than anecdotal).

P-A-Hs We've now nailed you:
Your structures on our walls.
How ironic
That your simplest
May be found in . . .
Old mothballs.

1. As it says above the simplest of the PAHs is the one used in old mothballs. What is it called?

   A. Naphthalene
   B. Anthracene
   C. Coronene

2. Which toxic element probably led to the phrase 'mad as a hatter?'

3. Which toxic element's Latin name forms the basis of the name of a trade linked with water?

4. 'Phossy jaw' was an agonizing and disfiguring disease caused by exposure to white phosphorus in the making of what?

5. '*Radium Girls*' is a 2018 film about the occupational exposure to radium while doing what?

6. Chronic mercury poisoning was attributed to which forensic procedure carried out by the Lancashire County Constabulary in the mid-twentieth century?

   A. Carbon-14 dating
   B. Drug testing
   C. Fingerprinting

7. Many people nowadays associate the acronym CBD with CBD oil, or cannabidiol oil (and CBD gummies) derived from the cannabis plant, legitimately taken by many to alleviate various complaints (and not having any of the psychoactive effects of other components of the plant such as THC, tetrahydrocannabinol). Yet CBD (chronic b____ disease) is also an acronym for a deadly occupational disease caused by work exposure to a particular element, featured in the sci-fi novel *Sucker Bait* by the US writer Isaac Asimov (1920–1992). What is the element?

8. Which naturally occurring fibrous mineral, used to strengthen and fireproof materials, has proved fatal to thousands due to work exposure and comes in several forms, including the white form (chrysotile), the brown form (amosite) and the blue form (crocidolite)?

9. Which inert gas is a major component of the gas mix that deep sea divers breathe?

One of the detectors at the Large Hadron Collider © D-VISIONS/Shutterstock

(or are they waves?)

THE DEATH-KNELL OF THE ATOM
Old Time is a-flying; the atoms are dying;
Come, list to their parting oration:-
We'll soon disappear to a heavenly sphere
On account of our disintegration.

Our action's spontaneous in atoms uranious
Or radious, actinious or thorious:
But for others, the gleam of a heaven-sent beam
Must encourage their efforts laborious.

For many a day we've been slipping away
While the savants still dozed in their slumbers;
Till at last came a man with gold-leaf and tin can
And detected our infinite numbers.

Thus the atoms in turn, we now clearly discern,
Fly to bits with the utmost facility;
They wend on their way, and in splitting, display
An absolute lack of stability.

Tis clear they should halt on the grave of old Dalton
On their path to celestial spheres;
And a few thousand million-let's say a quadrillion-
Should bedew it with reverent tears.

There's nothing facetious in the way that Lucretius
Imagined the Chaos to quiver;
And electrons to blunder, together, asunder,
In building up atoms for ever!

Sir William Ramsay (Discover of the noble gases, writing in *Nature, 1905*. Otto Hahn worked with Ramsay 1904–5 and was subsequently awarded the 1944 Nobel Prize in Chemistry for the discovery of nuclear fission)

'Arcane' is a word often levelled at the old art of alchemy. Arcane, perhaps too, the naming of some of the subatomic particles. Where will it all end and how much smaller can we probe? The irony is the smaller we want to go, the larger our colliders (spanning two countries in the case of the Large Hadron Collider), and the larger the budgets.

For those who don't know their muon from their gluon, a great introduction to the world of particle physics is Susie Sheehy's *The Matter of Everything: Twelve Experiments that Changed our World*, Bloomsbury Publishing, London (2022).

1. What was the first subatomic particle to be discovered?

2. *Flavour* is the name particle physicists give to different versions of the subatomic particles called *quarks*, which come in six different flavours: up, down, top, bottom, strange and what other flavour?

A. Cheese
B. Charm
C. Magic

3. *Particle Fever* was a US documentary film tracking experiments at the LHC (Large Hadron Collider) near Geneva from 2008–2012, and was one of the inaugural recipients in 2016 of the *Stephen Hawking Medal for Scientific Communication*. The documentary culminates with the successful identification of a subatomic particle predicted almost 40 years previous by and named after a British Nobel Laureate and theoretical physicist. What is the subatomic particle?

4. Unravel the names of these subatomic particles from their anagrams:

A. Her pony
B. Innate plot
C. No bows
D. No uncle

5. The following puzzle contains the names of 4 subatomic particles, 3 in the full circles and 1 in the intersections. See if you can find them:

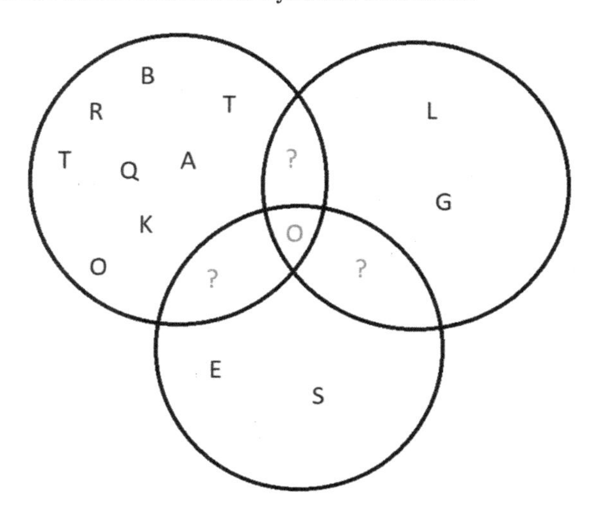

6. And again:

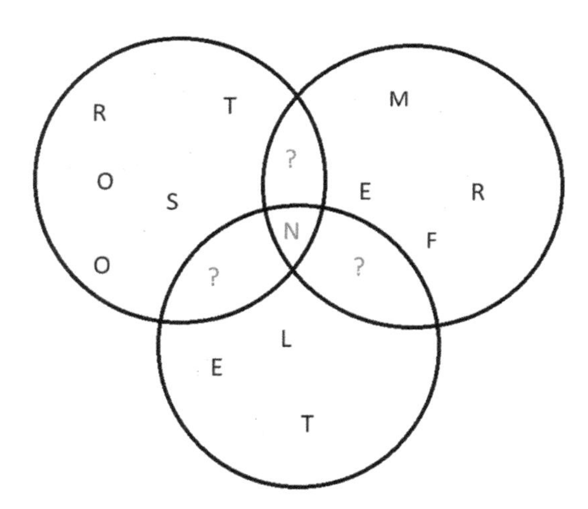

7. Which element is named after an Italian nuclear physicist who also has a subatomic particle named after him?

8. Which subatomic particles is the Irish playwright George Bernard Shaw (1856–1950) writing about here:

...why the men who believe in [subatomic particles] should regard themselves as less credulous than the men who believed in angels is not apparent to me.

9. Acrostic. Solve the clues across in the left-hand grid and reading down in the first column you will find the name of a subatomic particle (two words). Transfer the letters in the left-hand grid into the right-hand grid, as specified, and you will find a quotation from a nuclear scientist (and his name).

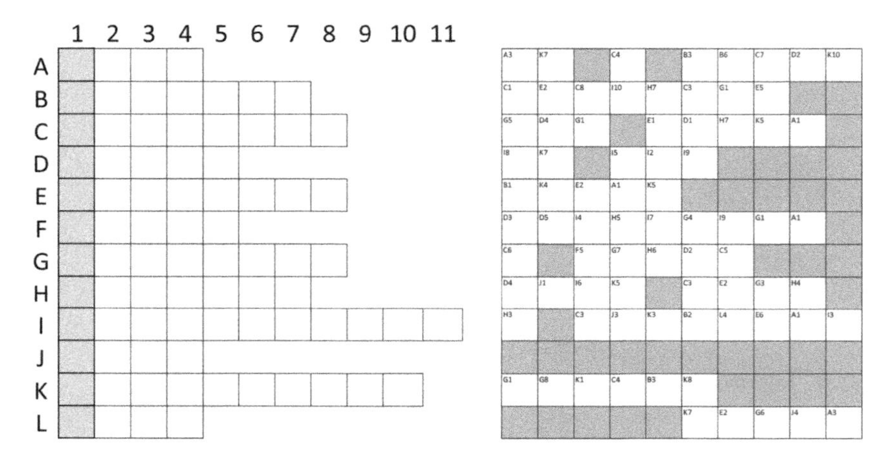

A. Angular momentum
B. Hypothetical particle that travels faster than light
C. Bose–Einstein condensation of atomic vapour was first seen in clouds of this element in 1995
D. _ _ _ _ _ particle, Helium-4 nucleus
E. The subject of the poem *Cosmic Gall* by John Updike (1932–2009)
F. Author of the 1961 book *The Atom and its Nucleus*
G. First subatomic particle to be discovered
H. _ _ _ _ _ _ _ mechanics
I. Section of the electromagnetic spectrum
J. Smallest unit of matter that forms a chemical element
K. New Zealand nuclear physicist with an element named after him
L. K meson

10. British physicist J J Thomson (1856–1940) discovered the electron and proposed the following model of the atom:

Reproduced from https://en.wikipedia.org/wiki/Plum_pudding_model#/media/File:PlumPuddingModel_ManyCorpuscles.png under the terms of the CC BY-SA license, https://creativecommons.org/licenses/by-sa/4.0/.

What dessert was the model named after?

A. Eton mess
B. Trifle
C. Plum pudding

11. CERN (Conseil Européen pour la Recherche Nucléaire) is the European Centre for Nuclear Research and is perhaps most famous for its Large Hadron Collider (LHC). Whilst working at CERN between 1989 and 1994, Tim Berners-Lee invented something the majority of us all now use. What was it?

12. The pronunciation of the name *quark* was chosen by the US theoretical physicist Murray Gell-Mann (1929–2019) after reading it in which of these works of fiction:

A. *The Hunting of the Snark* by Lewis Carroll
B. *The Owl and the Pussycat* by Edward Lear
C. *Finnegan's Wake* by James Joyce

13. Which particle accelerator did the inventor call a 'proton merry-go-round' and who was the inventor?

14. Muography is a technique similar to X-ray radiography but using muons. In 2017 this technique was used to discover a previously unknown void in which ancient large structure?

A. The Great Pyramid of Giza in Egypt
B. Lascaux Caves, France
C. Ellora Caves, India

15. Marietta Blau (1894–1970) was an Austrian physicist. Her photographic nuclear emulsions recorded tracks of subatomic particles and although she was nominated for the Nobel Prize several times, for both Physics and Chemistry, she never won it. However British physicist Cecil Powell (1903–1969) did win the 1950 Nobel Prize in Physics for the discovery of a subatomic particle using a development of Blau's photographic method. What particle?

A. Proton
B. Positron
C. Pion

16. Richard Feynman (1918–88), Julian Schwinger (1918–94) and Sin-Itiro Tomonaga (1906–79) were awarded the 1965 Nobel Prize in Physics for their work on QED. In this case, QED stands for Quantum Electrodynamics, the theory which describes the particle and wave nature of light. What else does QED stand for?

17. Which subatomic particle was predicted by Wolfgang Pauli, named by Enrico Fermi, and discovered by a team working on a project initially called *Project Poltergeist?*

A. Neutron
B. Neutrino
C. Fermion

18. You can find Pet Scanners at the vet's (they scan a microchip implanted under the pet's skin for the pet's identity and sometimes temperature). However, the PET Scanner was one of the many medical spin-offs that came from the perhaps more academic and esoteric work of particle physicists, and may be found in human hospitals. What do the letters in PET stand for?

A. Positron Emission Tomography
B. Proton Emission Tube
C. Particle Emission Testing

19. What do the Large Hadron Collider and the writer Lewis Carroll have in common (apart from having the initials LC)?

A. Alice
B. Mad Hatter
C. Jabberwocky

20. The Big European Bubble Chamber (BEBC) is a large detector previously used to study particle physics at CERN. It was filled with liquid hydrogen, liquid deuterium or a neon-hydrogen mixture. The bubble chamber was originally invented in 1952 by US physicist and neurobiologist Donald Glaser (1926–2013) for which he was awarded the 1960 Nobel Prize in Physics. What common beverage did Glaser use to fill his bubble chamber prototypes?

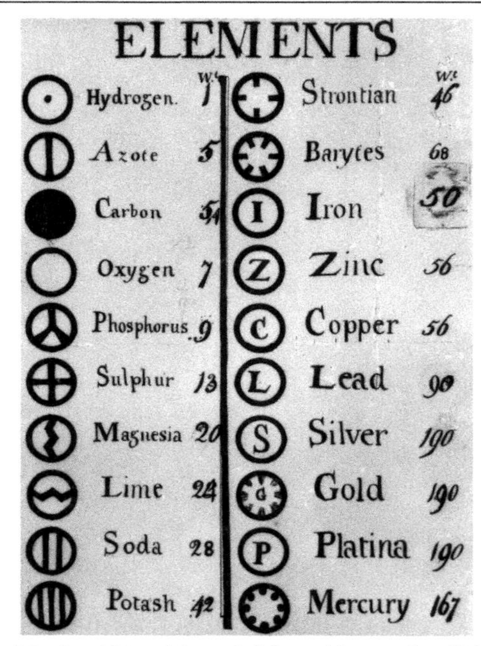

ELEMENTS

Element	Wt	Element	Wt
Hydrogen.	1	Strontian	46
Azote	5	Barytes	68
Carbon	54	Iron	50
Oxygen	7	Zinc	56
Phosphorus	9	Copper	56
Sulphur	13	Lead	90
Magnesia	20	Silver	190
Lime	24	Gold	190
Soda	28	Platina	190
Potash	42	Mercury	167

John Dalton's table of atomic weights and the symbols he used for a number of "elements". Compiled in 1808, some of the 20 substances included in the table are compounds and not pure elements: lime, for example. © Bettmann Archive/Getty Images

In 1866, the British chemist John Newlands presented his 'Law of Octaves' to a meeting of the Chemical Society, saying that if the elements were ordered according to their atomic weights, then 'the eighth element starting from a given one is a kind of repetition of the first, like the eighth note in an octave of music.' You only need to look at the alkali metals (Group 1) and the halogens (Group 17) to understand this concept and it was one of the first times when periodicity of the elements was described.

Yet Newlands was derided for his idea with one critic asking if he had tried an alphabetical arrangement. In a few years, Newlands' concept of periodicity was refined and defined by the chemical cartographer and polymath Dmitri Mendeleev into what we now know as the Periodic Table, cleverly leaving gaps for elements undiscovered at the time. In 1955, a new element, number 101 was named in his honour (Newlands did not achieve such fame, not even being elected to the Fellowship of the Royal Society, but the Royal Society did award him the Davy Medal for his efforts, named of course after the Cornish chemist Sir Humphry Davy without whose earlier electrochemical experiments, both Newlands and Mendeleev would have had far fewer elements to work with).

Over the years, as more and more elements have been discovered (or synthesised, albeit for the tiniest fractions of a second), taking the 63 in Mendeleev's Periodic Table to 118 (at time of publication), there have been claims and counterclaims for precedence and the right to name an element, often with IUPAC (The International Union of Pure and Applied Chemistry) or IUPAP (The International Union of Pure and Applied Physics) being the ultimate referees.

1. Lead was traditionally used for fishing weights but is now frowned upon due to its toxicity. Which element whose name comes from the Swedish for 'heavy stone' is preferred?

2. Which element's name is derived from the Ancient Greek for lead (but is not lead)?

3. Which two elements are named after their smelliness?

4. Which gaseous element was once called 'dephlogisticated air?'

   A. Hydrogen
   B. Nitrogen
   C. Oxygen

5. Which two elements have all five vowels in their names, and each vowel appearing once only?

6. Dmitri Mendeleev not only left gaps in his Periodic Table for elements yet to be discovered, he had the temerity to predict some of their chemical properties. For his predicted elements, he used the prefixes *eka*, *dvi* and *tri* (Sanskrit for one, two and three respectively). In 1871 he predicted that eka-boron (Eb) would have an atomic weight of 44, form an oxide with the formula Eb_2O_3, a sulfate with the formula $Eb_2(SO_4)_3$, and have a specific gravity of 3.5. Eka-boron was discovered in 1879 and matched up to his predictions with an atomic weight (measured then) of 43.79, the oxide and sulfate exactly matching the formulae and a specific gravity of 3.86. What is the element's name today?

7. Below 13.2 °C, this element's metallic β-form transforms itself into the semimetallic, crumbly α-form. Once a tiny portion is so 'infected' this so-called plague attacks the surrounding material. For this reason, Napoleon's army stood literally in its shirtsleeves in Moscow in 1812, since hundreds of thousands of uniform buttons disintegrated into powder. What is the element?

   A. Nickel
   B. Tin
   C. Zinc

8. John Venn (1834–1923) was an inventor (or perhaps inVenntor?). For instance, he invented a machine to bowl cricket balls. But he was also a philosopher and mathematician, remembered by many for his Venn Diagrams on which this puzzle is (very loosely) based.

For each of the puzzles below, replace the 3 question marks with letters to form an element from the four letters in the middle intersections, so that the letters in each of the 3 full circles also spells out an element. What are the four elements in each case?

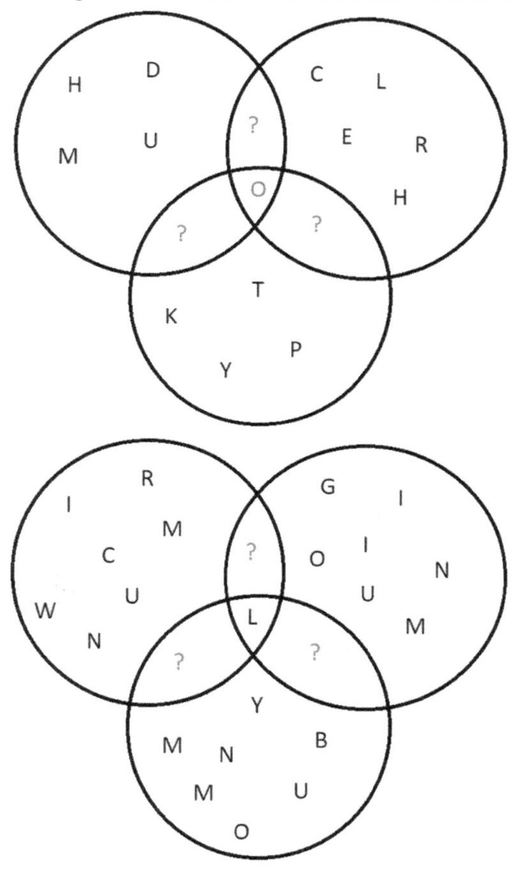

9. There are many alternative renditions of the Periodic Table, such as the one below. See if you can fill in the elemental symbols left out as indicated by the red circles. Then, taking these four elemental symbols (but not switching the letters in two-letter symbols, rearrange these to form a fifth element. What are the five elements?

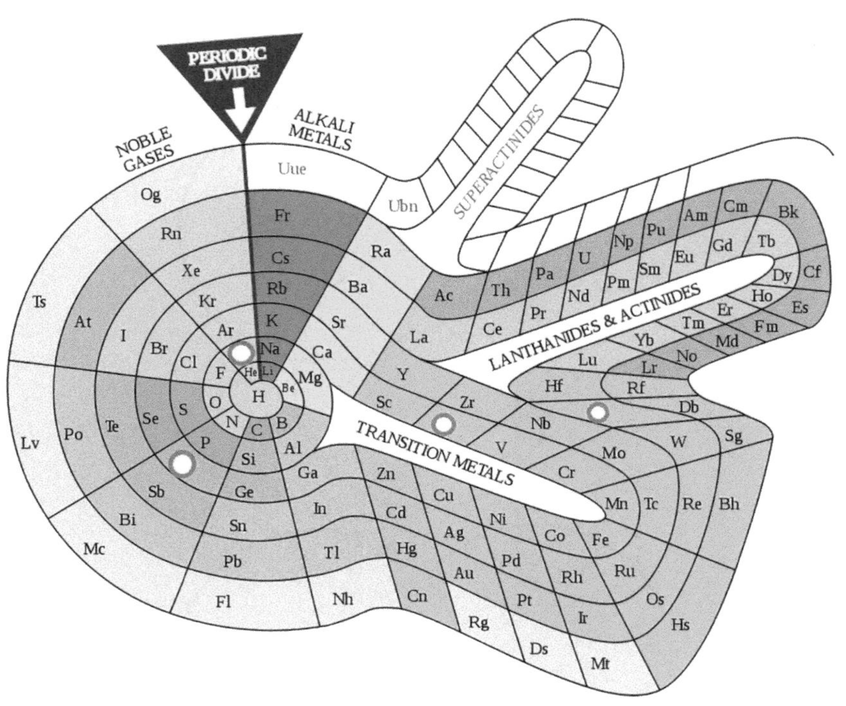

Adapted from https://en.wikipedia.org/wiki/Alternative_periodic_tables#/media/File:Elementspiral_(polyatomic).svg, under the terms of the CC BY-SA 3.0 license, https://creativecommons.org/licenses/by-sa/3.0/.

10. The elemental symbols of the precious metals gold and silver, Au and Ag are derived from their Latin names *aurum* and *argentum* respectively. Which other element actually contains the word 'Latin' in its (English) name and is also a precious metal?

11. *This valuable metal possesses the whiteness of silver, the indestructibility of gold, the tenacity of iron, the fusibility of copper, the lightness of glass. It is easily wrought, is very widely distributed, forming the base of most of the rocks, is three times lighter than iron, and seems to have been created for the express purpose of furnishing us with the material for our projectile.*

Jules Verne (1828–1905), French writer, in *From Earth to the Moon,* describing a metallic element to be used for a manned trip to the moon. What was the element?

12. Which metallic element (not mercury) has such a low melting point that it can melt in the palm of your hand?

13. Apart from aluminium (or aluminum), which other element has a different spelling in American English?

14. Which element is the author writing about here:

MRS. B.

...There is...a beautiful green salt too curious to be omitted; it is produced by the combination of______ with muriatic acid, which has the singular property of forming what is called sympathetic ink. Characters written with this solution are invisible when cold, but when a gentle heat is applied, they assume a fine blueish green colour.

CAROLINE.

I think one might draw very curious landscapes with the assistance of this ink. I would first make a water-colour drawing of a winter scene, in which the trees would be leafless, and the grass scarcely green; I would then trace all the verdure with the invisible ink, and whenever I chose to create spring, I should hold it before the fire, and its warmth would cover the landscape with a rich foliage.

MRS. B.

That would be a very amusing experiment, and I advise you by all means to try it.

Jane Marcet (1769-1858) in *Conversations on Chemistry in Which the Elements of that Science are Familiarly Explained and Illustrated by Experiments* (Volume 1 (of 2) Longman & Co 1825).

[muriatic acid is now better known as hydrochloric acid, HCl].

15. Threads of which metallic element are woven into gloves so that they can be used to operate touch-screen devices?

16. A precious metal used for crucibles, due to its high melting point (1772 °C) and resistance to oxidation, this element was particularly fashionable for jewellery during 'La Belle Epoque', the period before World War I. The USA's first coin made with it was minted in 1997, but it came down a peg or two when polymer chemist Hermann Mark fled Austria in 1938, carrying his fortune in the form of coat hangers made from the metal (and his wife knitted covers). What is the element?

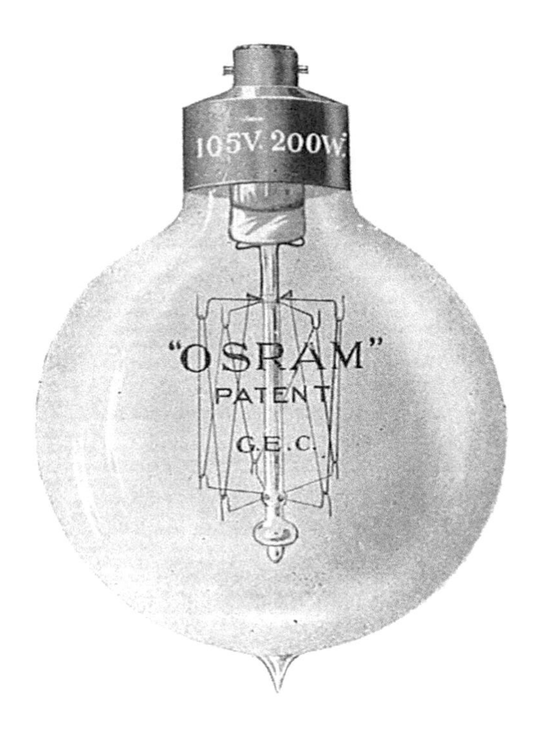

17. The OSRAM incandescent light (above) was developed in 1906 by the Austrian scientist and inventor Carl Auer von Welsbach (1858–1929). He also separated *didymium* into the two elements *neodymium* and *praseodymium* in 1885 and was one of three scientists to independently isolate *lutetium* (which he named *cassiopeium*).

OSRAM (or Osram Licht AG) is a German company that still manufactures electric lights. The word OSRAM combines the name of two elements which were being used for light filaments when the company was founded. What are the two elements?

18. Elemental crossword
Each crossword square can have one or two letters, which must be letters of an elemental symbol. (The Ds and Ts already in the grid stand for deuterium and tritium, isotopes of hydrogen). For example, Dalton could be written D(Al)TON.

Across

1. 7 Across, 28 Across Rule of physics which states that whenever two objects interact, they exert equal and opposite forces on each other (7, 5, 3, 2, 6)
7. See 1 Across (4, 2)
9. Designs (8)
10. Sausages (10)
11. Plant attractive to felines (6)
12. Adjoin (12)
14. British chemist known for his Law of Octaves concerning the periodicity of the elements (8)
15. Laboratory glassware (4, 5)
17. Nuclear _______, subdivision of a nucleus into two or more smaller nuclei (7)
19. Colloquial term for talking for too long (6, 2)
22. Containing silver (13)
24. Central themes (6)
25. Printing of written or visual material (11)
27. _______ of Seville, Spanish scholar and cleric who compiled an etymological encyclopaedia and invented the full stop, comma and colon (7)
28. See 1 Across (6)
29. Determination (13)

Down

2. Electrical circuit used to measure an unknown resistance, invented by British physicist and mathematician Samuel Hunter Christie, but named after another scientist and inventor who improved it (10, 6)
3. Crams with too many people (13)
4. Uninterrupted (9)
5. Axes (8)
6. Dries (10)
7. Wool wax (7)
8. Meteorological measure of cloud cover (4)
13. Geological period (13)
16. Tributes (12)
18. Taking place away from prepared ski runs (3–5)
20. Japanese moral code ('the way of the warrior') (7)
21. The _ _ _ _ _ _ Garden, a set of two poems and one of the first popular science books by the poet and naturalist Erasmus Darwin (grandfather of Charles Darwin) (7)
23. Alarm bell (6)
26. Type of curry (5)

19. Name these elements from their anagrams:

A. Atomic turnip
B. Like umber
C. Live forum
D. My tribute
E. Warm nuclei
F. Francium oil
G. Nag seamen
H. Dreamy opiums
I. India mogul

20. Name the elements (with number of letters provided) which may be spelt by rearranging some of the letters of the following elements:

A. Americium (6)
B. Californium – 2 elements (4, 8)
C. Dysprosium (6)
D. Gadolinium – 2 elements (4, 6)
E. Praseodymium – 3 elements (6, 6, 6)
F. Protactinium – 3 elements (3, 8, 8)
G. Roentgenium – 4 elements (3, 4, 4, 8)
H. Rutherfordium – 2 elements (7, 7)
I. Seaborgium – 2 elements (6, 6)
J. Xenon (4)
K. Ytterbium – 3 elements (7, 7, 6)
L. Zirconium – 2 elements (4, 4)

21. Use the elemental symbols on the right (only once) to complete the word grid below. One of the words is the surname of a German chemist who lends his name to a common piece of laboratory equipment.

		N		N
I	■	I	■	
S				S
	■	He	■	He
	I	S		S

B	C
Ho	K
O	P
Se	Ta
Te	U

22. Which synthetic element was named after the Russian chemist and originator of the Periodic Table?

23. Insert a different metallic element into each of the gaps below to make two phrases for each element:

A. Sterling ____________ Spoon
B. Cast ____________ Lady
C. Fool's ____________ Standard
D. Red ____________ Weight

24. Similar to a word ladder, each answer below provides letters that must appear in the answer immediately following it. The puzzle's final answer is an element.

A. _ _ _	Longest portion of geologic time (3)
B. _ _ _ _	Inert gas (4)
C. _ _ _ _ _	Type of saw (5)
D. _ _ _ _ _ _ _	Subatomic particle (7)
E. _ _ _ _ _ _ _ _	Subatomic particle (8)
F. _ _ _ _ _ _ _ _ _ _ _	Element (11)

25. Link these elements to their electron configurations:

1. Caesium	**A.** $[Xe]4f^15d^16S^2$
2. Calcium	**B.** $[He]2s^22P^2$
3. Carbon	**C.** $[Xe]6S^1$
4. Cerium	**D.** $[Ar]4S^2$

26. Elemental anagrams with a difference – they also contain the element's symbol in the anagram (for instance, fluorine F can make the anagram 'fine flour'). See if you can unpick these:

A. Clinch role
B. Nice link
C. Ham sushis
D. I cut maniac
E. Cram ruff in
F. Mind you mend
G. Timber tub

27. These are different forms (known as *allotropes*) of the same element. What is the element and what are the allotrope names?

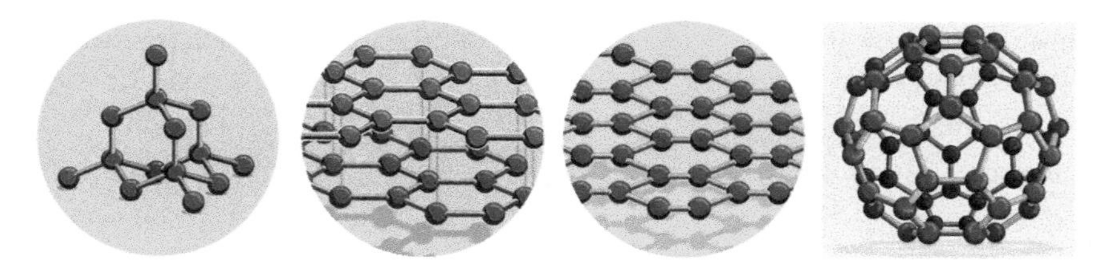

28. Which element is the author writing about here, the 'light' being a phenomenon which some schoolchildren have seen in their chemistry classes when a ribbon of the metal has been set fire to:

_ _ _ _ _ _ _ _ _, *a metal that yields light much like Sun,*
Being rich in chemically active rays,
Is a soft silver-white metal, that fuses
At low heat, and can be distilled at red heat.

J Carrington Sellars in *Chemistianity (Popular Knowledge of Chemistry) A Poem; also an Oratorical Verse on Each Known Element in the Universe* (1873).†

29. Time to look again at the Periodic Table. The following partial words may be completed by adding an elemental symbol at the front, and at the back, the symbol of the element immediately following the first horizontally in the Periodic Table, for instance:

_ _LLO_	(6)	Fat

Is TALLOW

73 Ta Tantalum	74 W Tungsten

A. _ASI_	(5)	Alkaline
B. _ARBO_	(6)	Element
C. _ _RBA_ _	(7)	Rubbish
D. _EUTRIN_	(8)	Subatomic particle
E. _LUORI_ _	(8)	Element
F. _ _TTI_ _	(7)	Crystal structure
G. _ _LEC_ _	(7)	Mammal
H. _ROTON_	(7)	Subatomic particles

† We know now of course that the light from the Sun is of a nuclear rather than a chemical origin.

30. Just like the question above, but this time the front and back inserts are the chemical symbols directly above and below each other in the Periodic Table, for instance:

_ _LA_ _	(6)	Animal fodder

Is SILAGE

A. _ _LM_ _	(6)	Fictional detective
B. _ CORSE_ _	(8)	US Film director
C. _ _AR_	(5)	Frightening
D. _ _RRA_ _	(7)	Notched
E. _ _ROI_ _	(7)	Courageous woman
F. _ _CT_ _	(6)	Ambrosia
G. _ _LENI_ _	(8)	Mineral
H. _ _WA_ _	(6)	Knight of the Round Table

31. *Tube Alloys* was the codename for the UK and Canada research and development programme for a nuclear deterrent during World War 2.

Link the following alloys to their constituent elements:

1. Brass	**A.** Carbon and iron
2. Bronze	**B.** Copper and zinc
3. Electrum	**C.** Copper and tin
4. Gunmetal	**D.** Copper, gold and silver
5. Nitinol	**E.** Copper, tin and zinc
6. Solder	**F.** Nickel and titanium
7. Steel	**G.** Tin and lead

32. Before the advent of the crossword in the early 20th century, the double acrostic was popular in the 19th century, and Queen Victoria was herself a fan.

Solve the clues below and take the first letter of each answer, reading down to form the name of an element. The last letter of each answer, reading up, will form the name of another element.

- Latin name for sodium, whence it gets its symbol (7)
- In the original place (2–4)
- Forest giraffe (5)
- British civil and mechanical engineer who built three ships, including the SS Great Britain (6)
- Unplanned (9)
- US State, the location of the Bryce Canyon National Park (4)
- Mathematician and pioneer of fractal geometry (10)

33. And the same rules for this sequence, again producing an element up and down, plus a third element from the first clue:

- Apart from berkelium, which synthetic element is named after a Californian city? (11)
- Taste characteristic of monosodium glutamate (5)
- Elementary particle, also a letter of the Greek alphabet (3)
- Swiss mathematician and polymath (5)
- US electrical engineer now made more famous by a brand of electric car (5)
- CHI_3 (8)
- $CO(NH_2)_2$ (4)
- Averages (5)

34. The below elements are known as the Lanthanides, after the first element in the series. All lanthanide elements fluoresce. Fluorescence occurs when light is shone on a material but the wavelength emitted back is different. Which of these elements (which fluoresces under UV light) is fittingly used in Euro banknotes to help prevent forgeries?

35. *Islands of Stability* describes a concept in Nuclear Physics where certain isotopes of superheavy elements have considerably longer half-lives, due to the stabilising effects of 'magic numbers' of protons and neutrons. These nuclides will appear in an island in the chart of nuclides.

Mathematics has another famous magic number, discovered by Indian mathematician Srinivasa Ramanujan (1887–1920). It is the smallest integer which may be expressed as the sum of the cubes of two different sets of positive integers. The number for these two sums of cubes will lead to that number:

$10^3 + 9^3$

$12^3 + 1^3$

What is the magic number?

36. Each six-letter answer should be written in a clockwise or anticlockwise direction around the relevant clue number. Once completed, the central grey hexagons will spell out an element. Two letters are already provided:

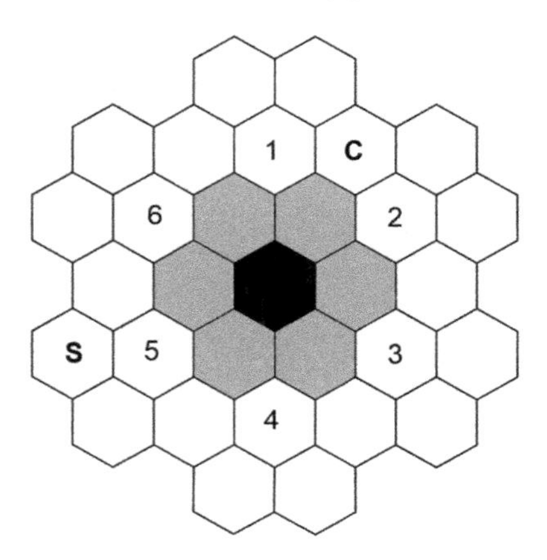

1. Not basic
2. Elongate cluster of flowers
3. Mustelid
4. Element known for its striking blue pigment
5. Gambling establishment
6. Acid with the formula HNO_3

37 Elemental Codeword Warm-Up. All cells must have elemental symbols in them as provided. A starting clue is that the number for carbon is its atomic number.

1	3	6	7	1
6		3		3
9	10	11	8	12
2		5		12
2	9	1	1	4

1	2	3	4	5	6
7	8	9	10	11	12
C	Er	F	K	N	O
P	Ra	S	U	W	Y

38. Elemental codeword. All elements with single-letter symbols are included and twelve of the others are given below the main grid to make an 'elemental alphabet' of 26. Your starting clues are that the numbers for hydrogen and oxygen match their atomic numbers. When you have completed the puzzle and filled in the 26 elemental letter grid you will also see the name of a solvent followed by the surnames of 3 scientists, and kitchen implements there. What is the solvent, who are the scientists and what are the kitchen implements?

5	8	2	24	3	7		14	7	24	21	14	7
3		21		23		7		16		14		21
21		20		21	9	1	4	21	23	9	5	19
8	9	21	8	9		21				21		23
17			3			20	8	7	12	5	16	
5	8	24	15	21	7	22		11			21	
	3			8				7			2	
	1			17		24	7	15	5	1	21	5
	7	23	5	5	3	8			8			8
6		1				13		10	8	7	8	9
14	18	21	7	21	8	9	6	15		8		5
25		9		8		26		25		17		1
7	16	21	9	9	15		24	7	15	5	1	19

1	2	3	4	5	6	7	8	9	10	11	12	13
14	15	16	17	18	19	20	21	22	23	24	25	26
B	C	F	H	I	K	N	O	P	S	U	V	W
Y	Ar	As	At	Ds	Er	Es	Ni	Re	Si	Ta	Te	Ts

39. See if you can match the English element names to the German:

1. Jod	**A.** Mercury
2. Quecksilber	**B.** Hydrogen
3. Sauerstoff	**C.** Iodine
4. Wasserstoff	**D.** Cerium
5. Zer	**E.** Oxygen

40. The Worshipful Company of Pewterers (pewterers.org.uk) is a 600-year-old traditional guild with a European Design Competition *Pewter Live* 'encouraging a contemporary take on the original craft of pewtersmithing.' Pewter is now usually an alloy of tin copper, bismuth, and antimony (with sometimes a little silver). What were the two main metals originally in pewter?

41. *Bleigiessen* is a German New Year tradition where families and friends melt a small amount of a certain metal in a spoon and pour it into cold water, where it solidifies. The shapes produced are then interpreted as an indication of what lies ahead in the coming year. For instance, if shaped like an elephant, someone will walk over you. If an owl, you need glasses. What is the metal?

42. What was the first man-made element?

A. Neptunium
B. Plutonium
C. Technetium

43. Marketed as 'The World's Most Wanted Pen' the Parker 51 was a fountain pen with nibs tipped in which metal?

A. Gold
B. Rhodium
C. Ruthenium

44. In 2017 IBM Research reported storing a single piece of data on a single atom of which element in *Nature*? The same element is used to produce a laser used in prostate surgery.

45. Which element plays hard to get?

46. Xe marks the spot. Fill the remaining cells in the grid below with the elemental symbols in the line below it to make four words, using each elemental symbol once and once only:

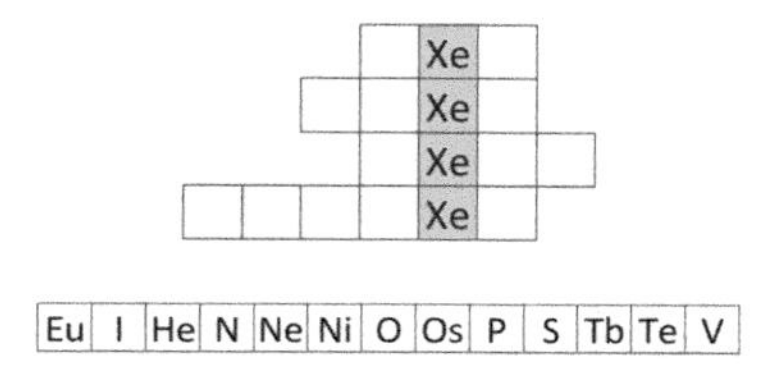

47. And finally. Before you leave this chapter, another puzzle like the one above, but using the last element‡ in the Periodic Table oganesson, Og.

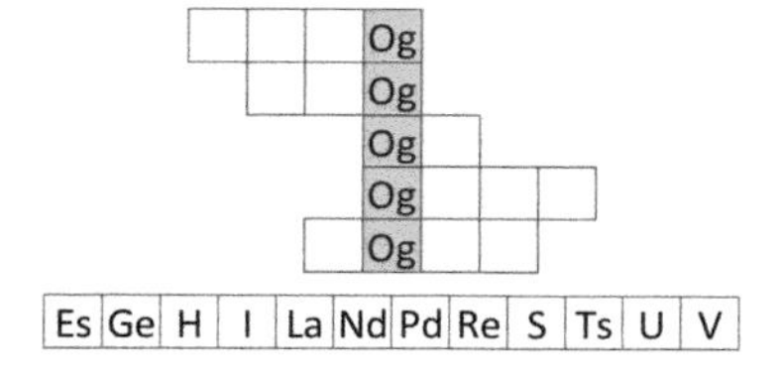

‡ As of 2022.

Young woman covered in coloured dye enjoying Holi festival in Jaipur, India. © Powerofforever/Getty Images

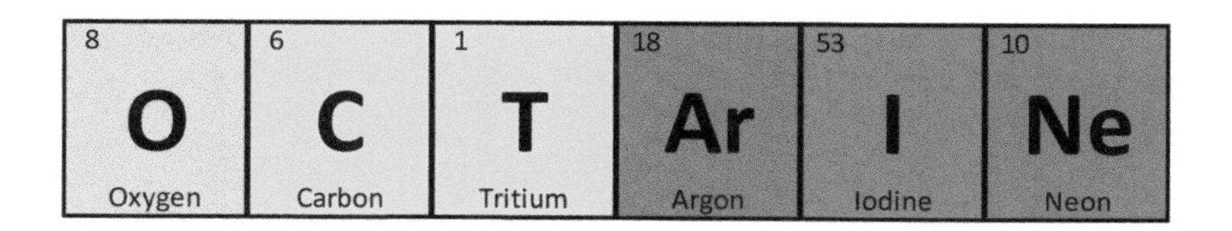

Pigments of the Imagination

Terry Pratchett (1948–2015) dreamt up a new colour *octarine* in his *Discworld* novels. But the world of colour hardly needs such imagination to stimulate, whether from the bold colours of mineral paint pigments, the amazing palette of nature and her plant-derived dyes, or artificial ones, particularly following the chance discovery by an 18-year old chemist (and painter) William Perkin, looking to synthesise the anti-malarial quinine from coal-tar in 1856. What he got instead was *mauve*, the first aniline dyestuff and the first commercialised synthetic dye.

50 years later Perkin crossed the pond to join chemistry's great and good for a celebratory dinner in New York (all diners sporting a specially made mauve bow tie rather than the formal black). By then there were 2000 or so artificial colours as a result of Perkin's work, and many other coal-tar derivatives too (such as artificial scents, food preservatives, photographic films, sweeteners and pharmaceuticals).

As Punch magazine said only 2 years after Perkin's discovery:

There's hardly a thing that a man can name
Of use or beauty in life's small game
But you can extract in alembic or jar
From the 'physical basis' of black coal-tar-
Oil and ointment, and wax and wine,
And the lovely colours called aniline;
You can make anything from a salve to a star
If you only know how, from black coal tar.

Not bad for what was basically a waste product from the production of town gas!

1. Which meteorological phenomenon does this mnemonic refer to: Richard Of York Gave Battle In Vain?

2. Chlorophyll gives the green colour to plants and is essential for photosynthesis. What is the name of the yellow pigment that may also be found in plants and in egg yolks?

3. What connects these elements: chlorine, indium, iodine, rhodium, rubidium?

4. Harris Tweed originally got its various colours from the local lichen found in the Scottish Islands. To enable these natural colours to bind to the wool, something called a *mordant* (from the Latin *mordere* 'to bite') was used, a common practice in dyeing. What was the readily available mordant used for the Harris Tweed (although no longer)?

5. In the puzzle below, replace the 3 question marks with letters to form a mordant from the four letters in the middle intersections, so that the letters in each of the 3 full circles also spells out a colour, dye or pigment. What is the mordant and what are the other compounds?

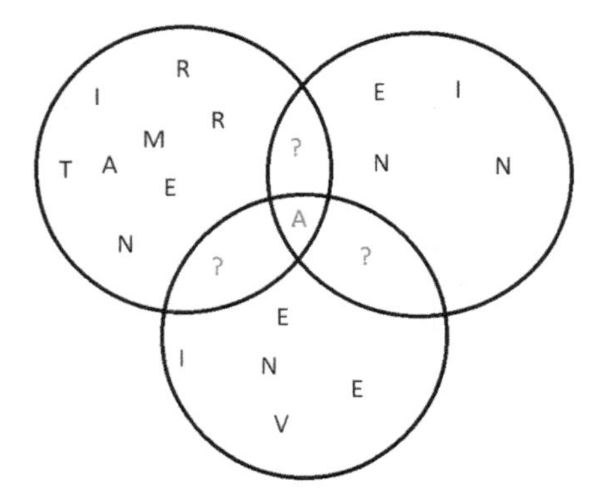

6. Woad is a plant that was used centuries ago to make a blue dye (and is still used for craft dyes). Can you turn WOAD into BLUE in 5 simple steps, one letter at a time, with each line being a bona fide word?

1. W O A D
2. _ _ _ _
3. _ _ _ _
4. _ _ _ _
5. _ _ _ _
6. _ _ _ _
7. B L U E

7. Van Gogh's most famous paintings are arguably his series of Sunflowers, yet the flowers have faded from a bright yellow to a dull brown since he painted them in 1888–89, due to the oxidation of a yellow pigment based on a metallic element. What is that element?

   A. Cadmium
   B. Chromium
   C. Cobalt

8. *The Man in the White Suit* was a 1951 Ealing comedy with Alec Guinness starring as the research chemist Sidney Stratton, who thinks he has invented a strong fibre which repels dirt and never wears out (which would mean catastrophe for the textiles and detergent industries). It was later reprised as a play at the Wyndham Theatre, London in 2019. The fabric is bright white.

   Which of these pigments is *not* white?

   A. Calcium carbonate
   B. Iron oxide
   C. Tin oxide
   D. Titanium dioxide
   E. Zinc oxide

9. Astaxanthin (named after the Latin for lobster *astacus*) is a pink carotenoid pigment found in algae and other microorganisms, crustaceans that feed on them, and fish such as salmon which feed on the crustaceans. What large bird owes its pink plumage to its diet of prawns and shrimps?

10. Pyrotechnic pigments. Link these firework colours to the elements that produce them:

1. Red	**A.** Barium
2. Orange	**B.** Calcium
3. Yellow	**C.** Copper
4. Green	**D.** Sodium
5. Blue	**E.** Strontium

11. Link these paint colours to their pigments:

1. Black	**A.** Cobalt(II) aluminate
2. Blue	**B.** Zinc chromate
3. Green	**C.** Carbon
4. Red	**D.** Iron(III) oxide
5. Yellow	**E.** Chromium(III) oxide

12. *Bhang* is a government-authorised preparation of a psychoactive drug. It is incorporated into spiced, milk-based drinks in the Indian city of Jaisalmer during Holi, 'Festival of Light and Colours' with festival goers throwing bright pigments and coloured water at each other. What is the drug?

13. *Vantablack* is a brand name for a class of super-black coatings invented by Ben Jensen and is claimed to be the be the 'world's darkest material' (and tell that to Philip Pullman!). What is its key ingredient?

A. Carbon black
B. Graphene
C. Carbon nanotubes

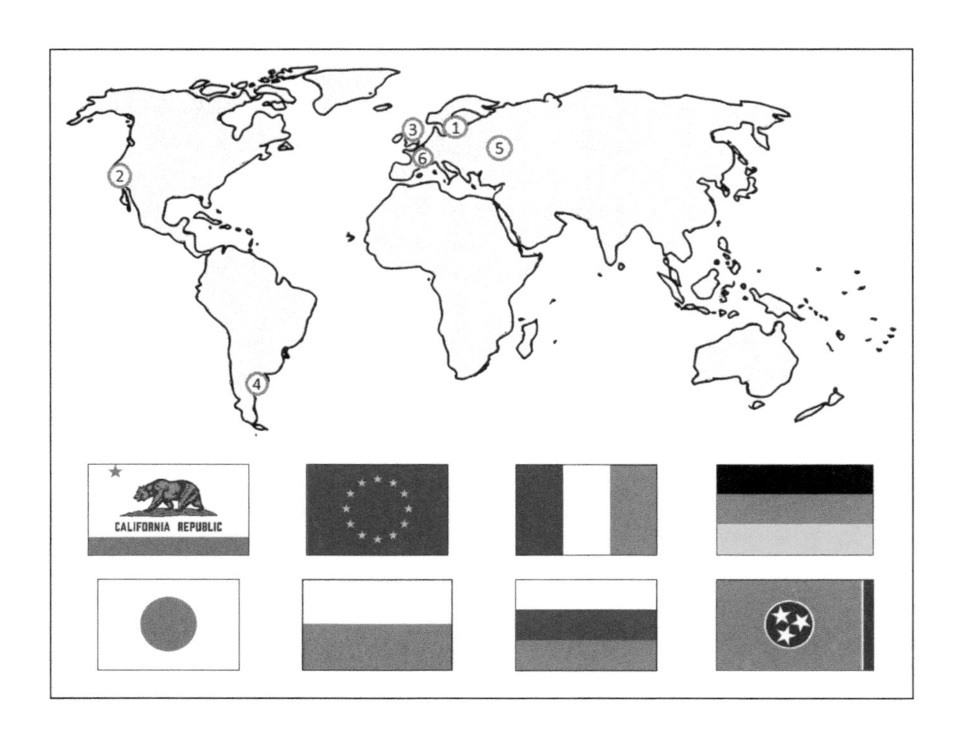
1
2
3
4
5
6
CALIFORNIA REPUBLIC

The Periodic Table is full of references to people but also many places, albeit sometimes in disguise, and elemental discoveries can be a source of international pride or international conflict. Indeed, the spat between American and Soviet scientists about the discovery (or synthesis) of the later elements has been dubbed (or should that be Dubna-ed?) by some nuclear scientists as the *Transfermium Wars*.

1. Connect the elements to the locations on the world map opposite:

   A. Berkelium
   B. Lutetium
   C. Moscovium
   D. Silver
   E. Strontium
   F. Ytterbium

   Which of the above elements is the odd one out?

2. What connects the eight flags on the opposite page?

3. Which element has a valley in Northern California named after it?

4. Which two elements are named after an area of Greece?

5. Which is the only element to be named after an Asian country?

6. Soufrière is a town on the Caribbean island of St Lucia. What element is it named after and why?

7. The gilding of the dome of the Cathedral of St Isaac at St Petersburg, Russia needed 100 kilograms of gold and reportedly claimed the lives of many men through associated poisoning, not with gold, but with a metallic element used to form an amalgam with gold to enable the gilding process. In the early nineteenth century, Birmingham's streets were not paved with gold, but running with this metal in the gutters surrounding the gilding factories. Using this gilding process, a gram of gold could gild 500 buttons: the buttons would be heated with the amalgam to drive the metal off, depositing the gold as a thin film. What was the metallic element?

8. What is the only element named after a river, and what is the river?

9. Which element's sodium salt is the name of a ghost town in Nevada that was named after its salt deposits? A medical dressing soaked in the element's acid is also Cockney rhyming slang for having no money. What is the slang and its rhyming counterpart?

Reproduced from https://en.wikipedia.org/wiki/Atomium#/media/File:Brussels_-_Atomium_2022.jpg under the terms of the CC BY-SA license, https://creativecommons.org/licenses/by-sa/4.0/.

10. The *Atomium*, pictured above is a popular tourist attraction in which European capital? It was constructed in 1958 to represent a unit cell of a crystal of an element? Which element?

11. Four elements are named after a single village in Sweden. Only one of these has a single letter for its elemental symbol. What is it, and what is the name of the village which also begins with the elemental symbol?

12. Which Welsh city was called 'Copperopolis' because of all the copper smelting done there?

A. Bangor
B. Cardiff
C. Swansea

13. Hafnium, holmium and lutetium are named after the Latin names for the capitals Copenhagen, Stockholm and Paris, respectively. Which element's (English) name contains the (English) name for another capital city?

14. Rare earth elements, important for many technological applications, are generally taken as the lanthanides. What country to date has been the predominant supplier?

A. Cameroon
B. Chile
C. China

15. Which element was originally named after a Mediterranean island?

16. Two of the elements are named after a single European country. What are the elements and what is the country?

17. Piccadilly Circus, London, is famous for its statue of Eros which was made in 1893 and was one of the first statues cast in a certain metal that was discovered earlier that century. What was the metal?

18. The Statue of Liberty† in New York has a green patina due to the weathering of the copper to form basic copper carbonate, commonly called *verdigris.* However, *verdigris* (which is also the name of a green pigment) gets its name from a European country. What is the country?

19. Gypsum, $CaSO_4{\cdot}2H_2O$, is a soft mineral widely used in the building trade. A name for the dehydrated mineral links it to a European capital which has large deposits in its quarries. What is the name?

† And as it so happens, rather fittingly 'Statue of Liberty' is an anagram of 'built to stay free'.

20. Similar to a word ladder, each answer below provides letters that must appear in the answer immediately following it. The puzzle's final answer is the London office of the Royal Society of Chemistry and also the Geological Society (the author has been to both on many occasions).

A. _ _ _	Alcoholic drink (3)
B. _ _ _ _	Chant (4)
C. _ _ _ _ _	Burn (5)
D. _ _ _ _ _ _	US inventor, actor and businessman known for his sewing machines (6)
E. _ _ _ _ _ _ _ _	British currency (8)
F. _ _ _ _ _ _ _ _ _ _ _ _ _ _ _	London offices (10, 5)

21. Despite his many outstanding contributions to science, the French chemist Antoine Lavoisier (1743–1794) never had an element named after him (likely if he did, its elemental symbol would have been Lv, which is now of course another element, livermorium). However, he did have an island named after him. Where is it?

A. Antarctica
B. Australia
C. Azores

22. Lavoisier is also recognised on the Eiffel Tower, along with 71 other French scientists, mathematicians and engineers (there are no women in this list).
Link these particular individuals so remembered with some of their histories:

1. André Marie Ampère (1775–1836)
2. George Cuvier (1769–1832)
3. Henri Victor Regnault (1810–1878)
4. Joseph Louis Gay-Lussac (1778–1850)
5. Lazare Carnot (1796–1832)
6. Léon Foucault (1819–68)
7. Pierre-Simon Laplace (1749–1827)

A. Chemist known for his hot air balloon ascents, he was partially blinded by a potassium explosion which demolished his laboratory, but he never lost his enthusiasm for experimentation. In addition to isolating and discovering elements and compounds, he pioneered the use of volumetric analysis
B. Founder of thermodynamics through his theory of an idealised heat engine, his father was Napoleon's Minister of War
C. Physical chemist and amateur photographer who was director of a porcelain laboratory. His laboratory was destroyed and his son killed during the Franco-Prussian War. He discovered several organochlorine compounds, including chloroform (CCl_4)
D. Physicist and inventor of the gyroscope, he measured the speed of light, and demonstrated the rotation of the earth with his pendulum
E. Physicist and mathematician who has the SI unit of electric current named after him
F. Polymath and assistant to Lavoisier who is best known for his work on celestial mechanics, he was Napoleon's examiner when Napoleon attended the *École Militaire*, Paris in 1784
G. Zoologist and anatomist, 'founding father of palaeontology' who strongly opposed theories of evolution

23. The English county of Cornwall is well known for its mining heritage, particularly copper and tin mining (and in 2006, UNESCO designated the Cornwall and West Devon Mining Landscape as a Word Heritage Site). A Cornish stream that is partially fed from mine run-off from some of these (now-abandoned) mines is called the *Red River* (or *Dowr Koner* in Cornish) due to the staining effect of pollutants. What metal gives rise to the red colour?

A. Copper
B. Iron
C. Tin

24. CERN's Large Hadron Collider is large enough to straddle two countries. Which countries?

25. A massive gas field was discovered in 2016 in the Tanzanian East African Rift Valley. Which gas?

26. Cheese Lane Shot Tower in Bristol, England is a Grade II listed building. Built in 1969, it was a replacement for an earlier shot tower, the first such tower ever built (in 1782 by William Watts). What was it designed to produce?

27. Architect Frank Gehry designed the Guggenheim Museum in Bilbao, Spain which opened in 1997. What metallic element is the building clad with?

28. St Paul's Cathedral was designed by Sir Christopher Wren after the Great Fire of London burnt the previous cathedral down in 1666. The domed roof is clad in lead. But what was Wren's original choice of metal for the roof?

29. Link these elementary American towns and cities with the States they may be found in:

1. Antimony	**A.** Alaska
2. Boron	**B.** Kentucky
3. Krypton	**C.** California
4. Mercury	**D.** Oklahoma
5. Platinum	**E.** Utah
6. Sulphur	**F.** Nevada

30. Which element was named after a country that didn't exist when its co-discoverer was born?

31. If Slovenia is silicon, Spain is einsteinium, and Mongolia is manganese, what is Austria?

A. Arsenic
B. Aluminium
C. Astatine

James Webb Space Telescope's main mirror assembly from the front with primary mirrors attached

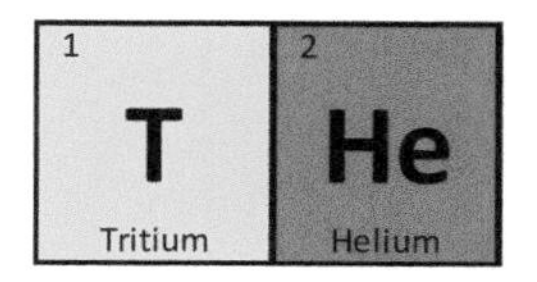

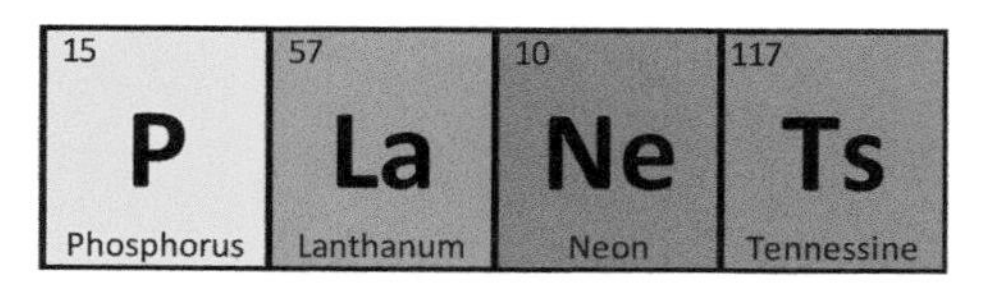

(and other astronomical stuff)

The word planet comes from the Greek meaning 'wanderer.' Let your mind wander around this section and enjoy! From the atomic to the astronomic, the heavens have revealed so much to those prepared to look. As new planets were being discovered, so too were new elements, and several of them were named after these new celestial bodies. Indeed, one element was originally discovered in the Sun and itself named after it.

1. Acrostic. Solve the clues across in the left-hand grid, and reading down in the first column you will find something astronomical (2 words). Transfer the letters in the left-hand grid into the right-hand grid as specified and you will find an appropriate quotation and the name of the person who wrote it.

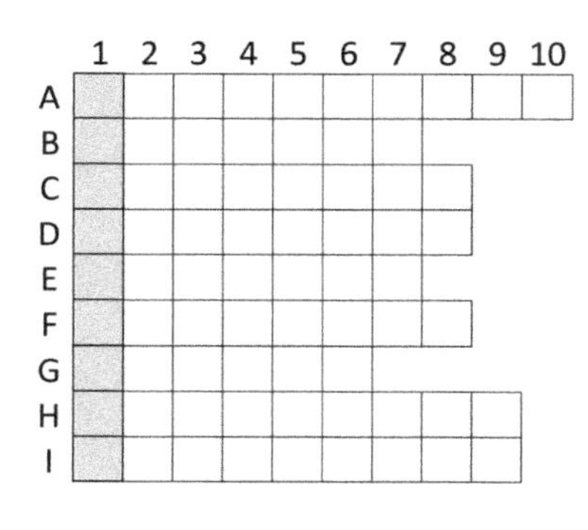

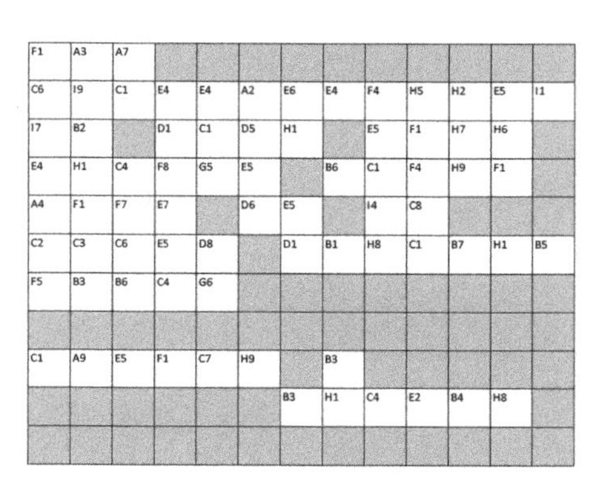

A. Substellar object not massive enough to sustain the normal nuclear fusion that fuels stars (2 words)
B. British astronomer who discovered helium in the Sun
C. Constellation of the zodiac
D. First name of one of the Herschel family of astronomers. She discovered eight new comets
E. Element that gives its name to a superhero's fictional planet birthplace
F. Major element of the Sun
G. Element which was named by the person in H below
H. French chemist with a minor planet named after him
I. British astrophysicist who discovered the mass-luminosity relationship of stars

2. Which two elements are named after the Earth and the Moon?

3. Which radioactive element was named after a planet but was later re-classified to a dwarf planet?

4. The following astronomical objects are named after women:

   A. _________ *Chain*, a group of galaxies in Cetus, named after the astronomer and astrophysicist Margaret _________ who held various leading positions including the Director of the Royal Greenwich observatory, President of the American Astronomical Society, and President of the American Association for the Advancement of Science.
   B. ______ Crater on Pluto, the name of the planet Pluto (now designated dwarf planet) that is credited as having been suggested by the girl Venetia ______, aged 11 after her grandfather read her the story of Pluto's discovery from *The Times* in 1930.
   C. A lunar crater is named after a Scottish astronomer who discovered the Horsehead Nebula in 1888. She shares her surname with the discoverer of penicillin (but no familial connection). What was her surname?
   D. She has craters on the Moon and Venus, and a main belt asteroid named after her, as well as an element. Albert Einstein called this physicist 'the German Marie Curie.' Who is she?

5. The following chemists have minor planets named after them. Fill in the blanks:

   A. D____I M___E___V (6, 9)
   B. __E__T___B__G (5, 1, 7)
   C. _V___E A__H__I_S (6, 9)
   D. _I__S P__L__G (5, 7)
   E. S_R H_M___Y _AV_ (3, 7, 4)
   F. J__N _A_T_N (4, 6)
   G. _O_E_T B_Y__ (6, 5)

6. An element discovered in 1669 was named after the Greek for the planet Venus (meaning light-bringer or light carrier). The Latin equivalent is *lucifer*, and indeed early matches made from this element were called lucifers. What is the element?

7. The Large Zenith Telescope at the Malcolm Knapp Research Forest at the University of British Columbia had a 6-metre diameter spinning liquid mirror. What was the liquid?

8. Which element, discovered in 1802, was named after the second asteroid discovered?

9.

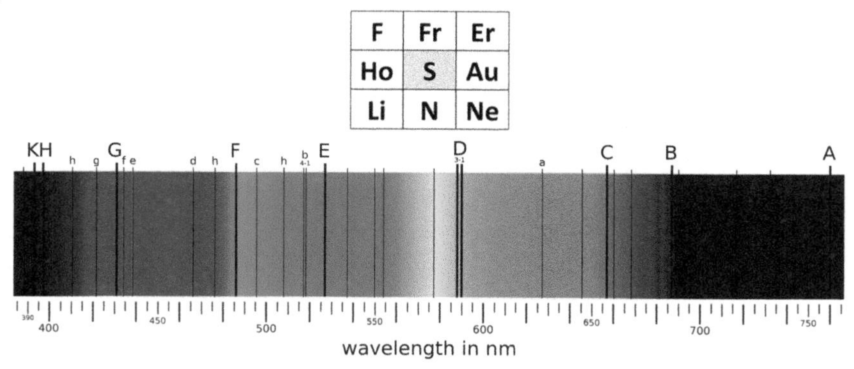

Using all 9 elemental symbols in the grid above (and not separating the letters of a two-letter symbol), spell out a solar 'spectral barcode' phenomenon shown above and named after a German physicist (two words).

Then answer these questions using some of the symbols always including the middle symbol and not switching letters with 2-letter symbols (using each symbol once and once only in each word):

A. Passenger ships (6)
B. Household fabric items (6)
C. Sharpens (5)
D. Plants (5)
E. Mythological creatures (5)

10. KRYPTON IS _ _ _ _ _ _ COME AND SEE FOR YOURSELF Signed CROOKES

(Sir) William Crookes (1832–1919) in a telegram to (Sir) William Ramsay (1852–1916) on examining the spectrum of a tube of 'krypton' and finding a yellow line that coincided exactly with the mysterious line of the Sun's spectrum. What is the element, named after the Greek word for the Sun?

11. Solve the anagrams below to reveal six celestial bodies (after removing one excess letter from each anagram). The excess letters reading down on the right will reveal a chemical element which was named after the first asteroid to be discovered:

- Clot up ____________ ___
- Heater ____________ ___
- Enter pun ____________ ___
- Deny image ____________ ___
- Emu curry ____________ ___
- Stun ram ____________ ___

12. Elementary Cryptogram.

Solve the puzzle in the cryptogram to find the stargazer and what is the element named after him? (hint – he was forward looking).

8	54	7	9	58	93	17
O	**Xe**	**N**	**F**	**Ce**	**Np**	**Cl**
Oxygen	Xenon	Nitrogen	Fluorine	Cerium	Neptunium	Chlorine

7	9	16	69	8	54	7	93	17
N	**F**	**S**	**Tm**	**O**	**Xe**	**N**	**Np**	**Cl**
Nitrogen	Fluorine	Sulfur	Thulium	Oxygen	Xenon	Nitrogen	Neptunium	Chlorine

13. B2FH? Not a compound but what is it?

A. An asteroid
B. A seminal paper on the stellar origin of the elements
C. A communications satellite

14. A founder of modern astrophysics Cecilia Payne (1900–79) struggled to get her work accepted. Prior to her work on the spectrum of the Sun, the commonly held view was that the Sun contained a lot of iron. She disproved this and showed which element predominated. What element is it?

15. Which noble gas is used in ion thrusters (also known as ion propulsion systems) for satellites and spacecraft?

16. Which astrophysicist wrote the following after it had been shown that most of the atom was empty space:

'When we compare the universe as it is now supposed to be with the universe as we had ordinarily preconceived it, the most arresting change is not the rearrangement of space and time by Einstein but the dissolution of all that we regard as most solid into tiny specks floating in void. That gives an abrupt jar to those who think that things are more or less what they seem. The revelation by modern physics of the void within the atom is more disturbing than the revelation by astronomy of the immense void of interstellar space.'

A. Annie Jump Cannon
B. Margaret Burbidge
C. Arthur Eddington

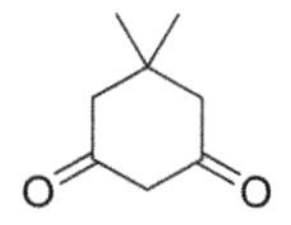

17. The above compound is dimedone, also known as methone (although the IUPAC name is 5,5-dimethycyclohexane-1,3-dione). It can be found in space. Where?

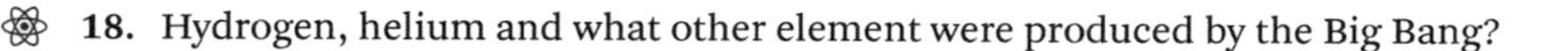

18. Hydrogen, helium and what other element were produced by the Big Bang?

19. The element neptunium, Np was discovered in 1940 by bombarding uranium with neutrons with a cyclotron at the Berkeley Radiation Laboratory in California. Neptunium follows uranium in the Periodic Table and was so named since Neptune is the next planet after Uranus in our Solar System. Yet this was not the first time an element had been named neptunium.

An element that had been predicted by Dmitri Mendeleev (1834–1907) from a gap in his Periodic Table (and which he called Ekasilicon, Es), was isolated in 1885 by the German chemist Clemens Winkler (1838–1904) who named it neptunium after the planet which had been discovered in 1846 (and had also been predicted previously before its discovery). What is the element's name now?

20. Costing a reported £8 billion to make, the James Webb Space Telescope was launched on 25 December 2021 and is intended to succeed the Hubble Telescope as NASA's flagship mission in astrophysics. In July 2022 its first deep field images were released. It is designed primarily for infrared astronomy and its primary mirror consists of 18 hexagonal mirror segments of which light metallic element coated with gold?

THE GREAT LOZENGE-MAKER.

A Hint to Paterfamilias.

An 1858 caricature by John Leech (1817–1864) for Punch magazine after the Bradford poisoning case

84	53	16	8	7	16
Po	**I**	**S**	**O**	**N**	**S**
Polonium	Iodine	Sulfur	Oxygen	Nitrogen	Sulfur

Mr Accum's account of water (i.e. the Companies' water-the filthy and unwholesome water supplied from the Thames, of which the delicate citizens of Westminster fill their tanks and stomachs, at the very spot where one hundred thousand cloacinae, containing every species of filth, and all unutterable things, and strongly impregnated with gas, the refuse and drainings of hospitals, slaughter-houses, colour, lead and soap works, drug-mills, manufacturers, and dung-hills, daily disgorge their abominable contents) is so fearful, that we see there is no wisdom in the well: and if we then fly to wine, we find, from his analysis, that there is no such truth in that liquid; bread turns out to be a crutch to help us onward to the grave, instead of this being the staff of life; in porter there is no support, in cordials no consolation; in almost everything poison, and in scarcely any medicine, cure!

From: Deadly Adulteration and Slow Poisoning; or Disease and Death in the Pot and the Bottle in which the Blood-Empoisoning and Life-Destroying Adulterations of Wines, Spirits, Beer, Bread, Flour, Tea, Sugar, Cheesemongery, Pastry, Confectionery, Medicines are Laid Open to the Public by an Enemy to Fraud and Villainy (*ca.* 1830).

What's your poison? A saying referring to one's choice of drink (usually alcoholic). And of course alcohol (ethanol) is a poison. The Swiss physician and alchemist Paracelsus (or rather more grandly Theophrastus Bombastus von Hohenheim – try saying that when you've had a few) may be paraphrased as saying 'The dose is the poison' and he is quite right. It is only a matter of amount that turns tonic into toxic.[†]

1. Which toxic element was implicated in a mass poisoning in 1858 in Bradford, England, in adulterated sweets sold by William Hardaker (known locally as *Humbug Willy*)? The same element also featured in a mass poisoning in the UK in 1901 due to the consumption of contaminated beer in the North of England.

A. Arsenic
B. Lead
C. Mercury

2. In 1985, some Austrian wines were found to be adulterated with diethylene glycol (to make the wines appear sweeter and more full-bodied). Diethylene glycol is an ingredient in what commonly used motoring product?

[†] For those wanting to read further, John Emsley has written some splendid books on poisons, both elemental and molecular, for which see the Bibliography.

3. Which toxic metal was implicated in the loss of all crew in John Franklin's ill-fated expedition (departing from England in 1845 in the *HMS Erebus* and the *HMS Terror*) to find the North-West Passage around the top of Canada, and implicated in the book and documentary film *Beethoven's Hair*?

4. Link these poisons with the novels they appear in (Spoiler Alert!):

   1. Arsenic
   2. Cantharidin (Spanish Fly)
   3. Cyanide
   4. Hyoscine
   5. Laburnum
   6. Nicotine
   7. Nitrobenzene
   8. Thallium

   A. *The Pale Horse* by Agatha Christie (1890–1976)
   B. *Death of a Ghost* by Margery Allingham (1904–1966)
   C. *Love in the Time of Cholera* by Gabriel García Márquez (1927–2014)
   D. *Madame Bovary* by Gustave Flaubert (1821–1880)
   E. *My Cousin Rachel* by Daphne du Maurier (1907–1989)
   F. *The Nursing Home Murder* by Ngaio Marsh (1895–1982)
   G. *The Poisoned Chocolates Case* by Anthony Berkeley (1893–1971)
   H. *My Uncle Oswald* by Roald Dahl (1916–1990)

5. Link these poisons with the famous or infamous (non-fiction) cases. Names include victims and perpetrators:

   1. Hemlock
   2. Hyoscine
   3. Polonium-210
   4. Ricin
   5. Sarin
   6. Strychnine

   A. Alexander Litvinenko
   B. Aum Shinrikyo cult (Tokyo subway)
   C. Dr Crippen
   D. Georgi Markov
   E. Thomas Neill Cream
   F. Socrates

6. Similar to a word ladder, each answer below provides letters that must appear in the answer immediately following it. The puzzle's final answer is a crime novel by Agatha Christie which (rather unusually) gives away the poison in the book's title.

A. _ _ _ — Semi-solid whose name was coined by the Scottish chemist Thomas Graham (3)

B. _ _ _ _ — Valley (4)

C. _ _ _ _ _ — Greek physician and philosopher in the Roman Empire (5)

D. _ _ _ _ _ _ — Lead ore (6)

E. _ _ _ _ _ _ _ _ _ — Painkiller (9)

F. _ _ _ _ _ _ _ _ _ _ _ _ _ _ _ _ — Crime novel (9, 7)

7. One German housewife failed in her murderous intent when her husband noticed that his spiked soup glowed in the dark. Thus forewarned, he didn't drink it, and a public analyst confirmed his suspicions that it contained a toxic element. What was it?

A. Phosphorus
B. Radium
C. Uranium

8. What poison connects the films *The Grand Budapest Hotel*, *Blueprint for Murder*, and *Psycho*?

A. Curare
B. Hemlock
C. Strychnine

9. Acrostic. Solve the clues across in the left-hand grid and reading down in the first column you will find another name for the plant mentioned in the last of the across clues. Transfer the letters in the left-hand grid into the right-hand grid as specified and you will find an appropriate quotation, the name of the person who wrote it and the work it's from.

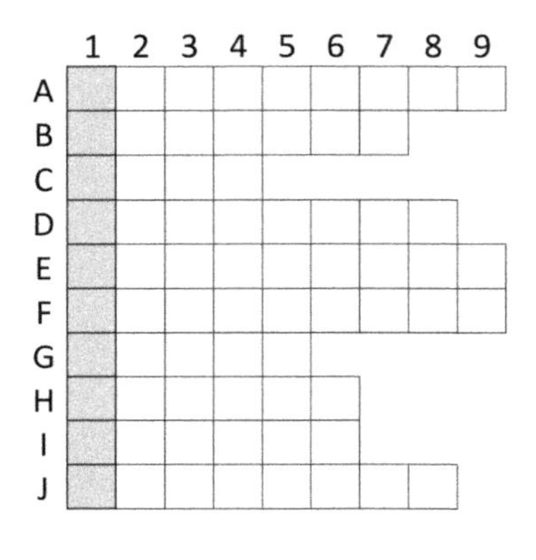

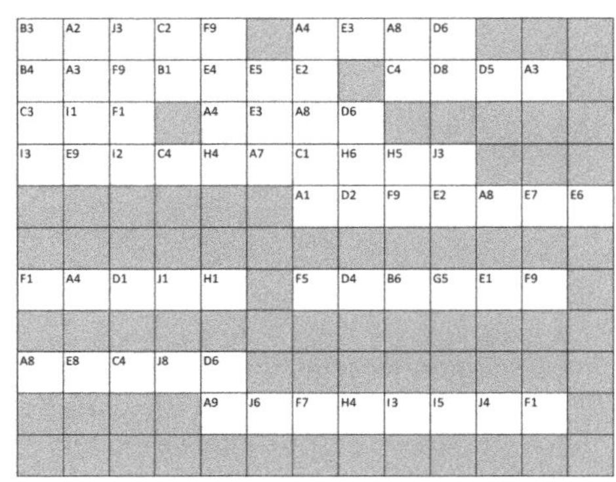

A. Toxic element featured in the sci-fi story *Sucker Bait* by Isaac Asimov (1920–1992)
B. C_2H_5OH
C. Metallic element in the compound in D below
D. Toxic compound once used to sweeten wines (old name)
E. Plant-derived toxin also known as wolfsbane and featured in the short story *Lord Arthur Savile's Crime* by Oscar Wilde (1854–1900)
F. Poison used in *The Unpleasantness at the Bellona Club* by Dorothy L Sayers (1893–1957)
G. Confessions of an English _ _ _ _ _ Eater, by Thomas De Quincey (1785–1859)
H. Common allergen due to its use in jewellery
I. Famous physicist who was exposed to high levels of mercury during his alchemical experiments
J. Alkaloid extracted from Deadly Nightshade and other plants

10. What alternative to table salt (for those who need a reduced sodium intake) can be lethal when injected (rather than taken orally) and indeed has been used to commit murder on numerous occasions and also to carry out the death penalty?

11. For some poisoners, a single poison was not enough. *Aqua Tofana* consisted of a lethal blend of arsenic, lead and belladonna and was created in Sicily around 1630. The following jumbles of letters contain binary mixtures of poisons. To make it slightly easier than an anagram, despite the letters being jumbled, letters are in the right order as they appear in the two words.

For instance BARELLASDOENINNAC unravels to reveal arsenic and belladonna [B] AR[E][L][L][A]S[D][O]E[N]I[N]N[A]C.

Pick apart these poisons:

A. MERCYANCUIRDYE
B. LSETRYCAHNINED
C. PRICOLONIUINM
D. CANADTIMMIOUNMY

12. In the 1995 film *A Young Poisoner's Handbook*, actor Hugh O'Connor portrays the serial killer Graham Young. What toxic element did Young use to poison his victims?

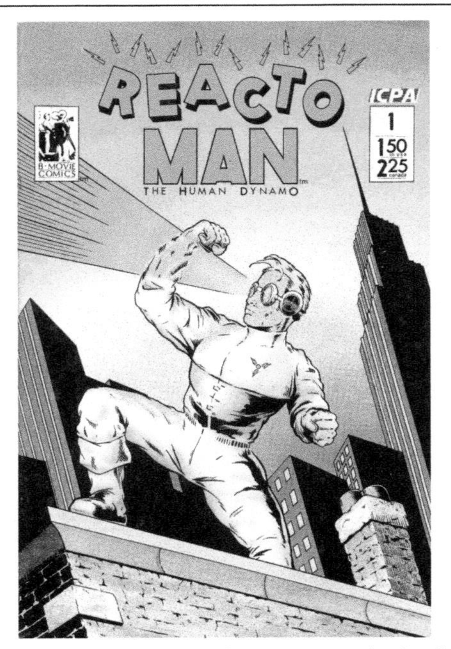

Research scientist Trevor Stratton turns into superhero Reactoman when irradiating radium crystals.
Image courtesy of mycomicshop.com

Although the term 'radioactivity' was coined by Marie Curie in 1898, it was originally Henri Becquerel who discovered this phenomenon two years earlier (in uranium salts by their action on a photographic plate). Both Henri Becquerel and Marie (and Pierre) Curie shared the 1903 Nobel Prize in Physics for their work and their names were subsequently immortalised in the lexicon of science.

By 1913, the sci-fi writer and futurist H. G. Wells predicted an atomic bomb in his novel *The World Set Free.* The physicist and a central figure in the Manhattan Project, Leo Szilard said the book made a great impression on him. In less than 50 years after Becquerel's serendipitous discovery, Wells' prescient prediction became all too horribly real with uranium-based and plutonium-based bombs being dropped on Hiroshima and Nagasaki, respectively.

1. Which Greek letters describe three forms of nuclear radiation?

2. Rosamund Pike played a pioneering radiochemist in the 2019 film *Radioactive.* Who was she?

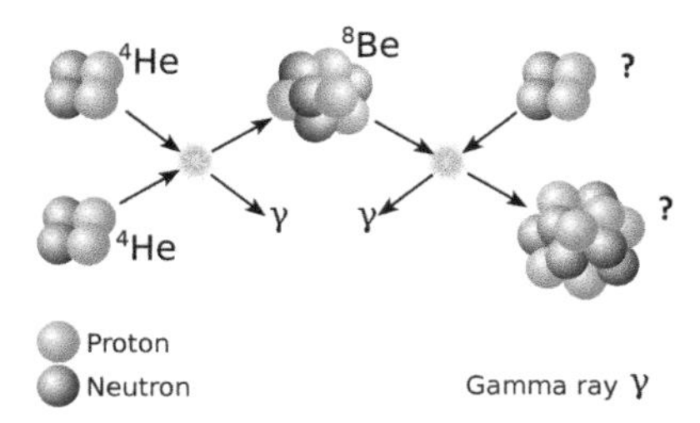

3. The *Triple-alpha process* illustrated above describes a set of stellar nuclear fusion reactions. What are the two end products?

4. What nuclear reaction is named after the process of cell division?

5. In 1941, the chemistry undergraduate named Margaret Melhase (later Margaret Melhase Fuchs) discovered a radioactive isotope which later proved very useful in medicine, geology and the measurement of radiation leaks (such as those from Chernobyl). Which of these isotopes was it?

   A. Caesium-137
   B. Iodine-131
   C. Uranium-235

6. The lightest radioactive element was also the first element to be produced artificially and its name comes from the Greek for artificial. What is it?

7. Radon is a radioactive noble gas which causes problems in geological hotspots, such as the county of Cornwall, England, necessitating special building measures to lower the radon dose in houses. On the continent, you can opt for a 'radon cure' at the five-star Hotel Radium Palace, with the deluxe suite being called the Marie Curie Suite. Where might this hotel be found?

   A. France
   B. Poland
   C. Czech Republic

8. John H. Lawrence (1904–1991) has been called the *Father of Nuclear Medicine.* Nuclear medicine employs radioactive tracers, radioactive sources and elementary particles to target and shrink tumours. His perhaps more famous brother invented a particle accelerator called the *cyclotron.* What was his name (and not just his surname!)?

9. Some may argue that this book is just a load of scientific trivia, the singular of which is *trivium.* But what is *tritium* and why might you find it on your wrist?

10. What radioactive element discovered in 1898 was Satan made of in the 1904 short story '*Sold to Satan*' by American writer Mark Twain (1835–1910)?

11. The promise of cheap energy seemed possible when two scientists, Stanley Pons (b 1943) and Martin Fleischmann (1927–2012) claimed they had achieved what (two words) in 1989, using electrolysis of heavy water (D_2O) and palladium (Pd)?

12. Which radioactive element may be found in a concentrate called 'Yellowcake' (but by no means eat it!)?

13. Which element used in the cladding of the Chernobyl nuclear reactor (which catastrophically had a meltdown in 1986) reacted with steam at high temperatures to form hydrogen and explode?

Rachel Carson © Roseed Abbas/Shutterstock. Emmanuelle Charpentier © ODD ANDERSEN/AFP via Getty Images. Marie Curie © Sheila Terry/Science Photo Library. Rosalind Franklin © Roseed Abbas/Shutterstock. August Kekulé © German Vizulis/Shutterstock. Stephanie Kwolek © Hagley Museum and Archive/ Science Photo Library. Antoine Lavoisier © ZU_09/Getty Images. James Lovelock © Anthony Howarth/Science Photo Library. Lise Meitner © Central Press/Getty Images. Dmitri Mendeleev © Gary Brown/Science Photo Library. Isaac Newton © Roseed Abbas/Shutterstock. Hideyo Noguchi © Yosuke Hasegawa/Getty Images/ iStockphoto. Marguerite Perry © Science Source/Science Photo Library. George Washington Carver © Gwen Shockey/Science Photo Library.

Scientists and Inventors

Scientists should be on tap but not on top.

Winston Churchill (1874–1965)

The polymath William Whewell coined the term 'scientist' in 1834, originally for the Scottish polymath Mary Somerville who has an Oxford University College named after her. But it is perhaps unlikely that Whewell could have foreseen 'Scientist' being the moniker of Hopeton Overton Brown, a Jamaican recording engineer and producer known for his dub music with albums including *Scientist encounters Pac-Man* and *Scientist meets the Space Invaders.*

(Before Whewell, scientists tended to be called *natural philosophers*).

1. Match the images opposite with the scientists:

A Dorothy Hodgkin, X-ray crystallographer of biomolecules
B Charles Darwin, author of *On the Origin of Species*
C Rosalind Franklin, X-ray crystallographer who helped elucidate the structure of DNA
D Antoine Lavoisier, French chemist who revolutionised chemistry but was a victim of the French revolution
E Rachel Carson, author of *Silent Spring* which alerted the world to the dangers of the misuse of pesticides
F George Washington Carver, American agricultural scientist
G Stephanie Kwolek, chemist who invented Kevlar®
H Sir Isaac Newton, alchemist and physicist known for his Laws of Motion
I Marguerite Perey, discoverer of the element Francium
J Dmitri Mendeleev, known for his Periodic Table
K Marie Curie, the only person to be awarded the Nobel Prize in two different scientific fields
L James Lovelock, inventor of the Electron Capture Detector
M Barbara McClintock, American cytogenecticist and Nobel Prize winner
N Alfred Nobel, inventor of dynamite and funder of the Nobel Prizes
O Lise Meitner who co-discovered nuclear fission
P Hideyo Noguchi, bacteriologist who discovered the causative agent of syphilis
Q Emmanuelle Charpentier, co-recipient of the 2020 Nobel Prize in Chemistry for the development of a method for genome editing

R Wilhelm Roentgen, dicsoverer of X-Rays
S Mary Anning, English palaeontologist
T August Kekulé who elucidated the ring structure of benzene (C_6H_6)

2. Link the scientists to the actors who portrayed them in the following films:

1. Sigourney Weaver in *Gorillas in the Mist* (1988)	**A.** Mary Anning
2. Daniel Craig in *Copenhagen* (2002)	**B.** Charles Darwin
3. David Bowie in *The Prestige* (2006)	**C.** Sir Alexander Fleming
4. Denis Lawson in *Breaking the Mould* (2009)	**D.** Dian Fossey
5. Paul Bettany in *Creation* (2009)	**E.** Stephen Hawking
6. Benedict Cumberbatch in *The Imitation Game* (2014)	**F.** Werner Heisenberg
7. Eddie Redmayne in *The Theory of Everything* (2014)	**G.** Nikola Tesla
8. Kate Winslet in *Ammonite* (2020)	**H.** Alan Turing

3. Which synthetic element was named after the discoverer of X-rays?

4. Which two elements are named after a Danish physicist and his birthplace respectively?

5. Which element led to a naming controversy before its final name was decided upon, originally being named after the physicist in the question above, or the German chemist Otto Hahn (1879–1968)?

6. And which element also had names proposed after Otto Hahn and French physicist Henri Becquerel (1852–1908) before its final name was decided upon?

7. Family connections:

   A Max Born was awarded the 1954 Nobel Prize in Physics for his work on quantum physics. His son-in-law (married to his daughter Irene Born) was a codebreaker at Bletchley Park during WWII. One of his son-in-law's daughters was a well-known singer. Who was she?

   1. Lulu
   2. Olivia Newton-John
   3. Madonna

   B Which famous scientist was the grandson of the potter Josiah Wedgwood (1730–1795)?

   C English biologist Thomas Henry Huxley (1825–1895) had a prominent family which included:

   - Grandson Sir Julian Huxley (1887–1975), English evolutionary biologist, first director of UNESCO and one of the founding members of the World Wildlife Fund.
   - Grandson Sir Andrew Fielding Huxley (1917–2012), English physiologist and biophysicist who shared the 1963 Nobel Prize in Physiology or Medicine for his work on nerve impulses.
   - Grandson, philosopher and writer, author of the dystopian novel *Brave New World.* He was nominated for the Nobel Prize in Literature nine times but never got it. **What was his name?** He also taught Eric Arthur Blair at Eton College. Eric Arthur Blair would also become famous for another dystopian novel under a pseudonym. **What was the pseudonym and this writer's dystopian novel?**

   D Which family has received the most Nobel Prizes?

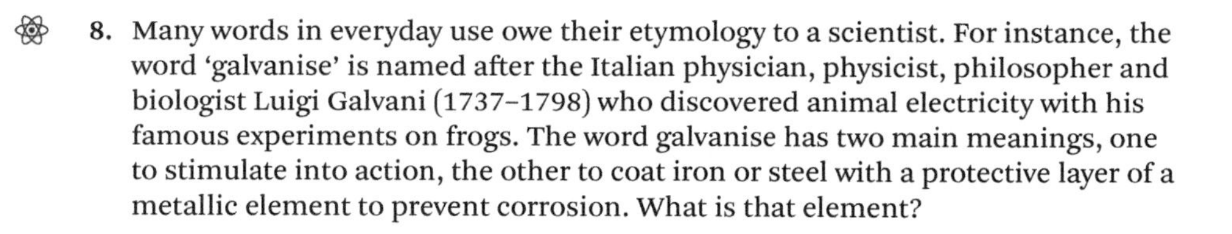

8. Many words in everyday use owe their etymology to a scientist. For instance, the word 'galvanise' is named after the Italian physician, physicist, philosopher and biologist Luigi Galvani (1737–1798) who discovered animal electricity with his famous experiments on frogs. The word galvanise has two main meanings, one to stimulate into action, the other to coat iron or steel with a protective layer of a metallic element to prevent corrosion. What is that element?

9. The cosmologist Stephen Hawking (1942–2018) wrote in his *A Brief History of Time* (1988):

Someone told me that each equation I included in the book would halve the sales.

Notwithstanding the above, link these famous equations to their authors:

1. $a^2 + b^2 = c^2$
2. $n\lambda = 2d \sin \theta$
3.

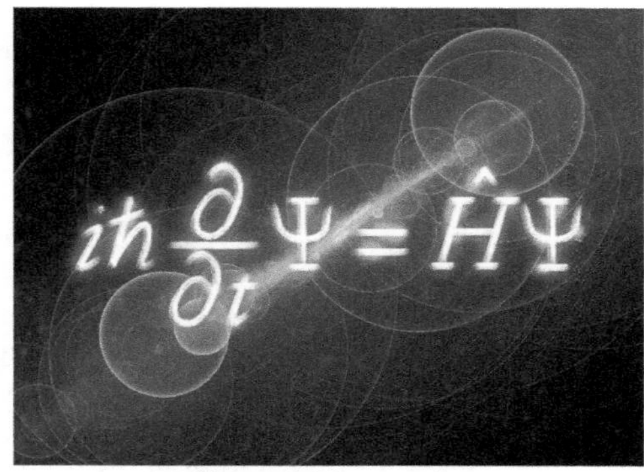

© *Sakkmesterke/Shutterstock*

4. $E = mc^2$
5. $F = Gm_1m_2/d^2$

a. Sir William Lawrence Bragg (1890–1971) and his father Sir William Henry Bragg (1862–1942)
b. Albert Einstein (1879–1955)
c. Sir Isaac Newton (1642–1727)
d. Erwin Schrödinger (1887–1961)
e. Pythagoras (c 560–480 BC)

10. What fact connects Ernest Rutherford, J. J. Thomson and Stephen Hawking?

A. They are related
B. They are all buried at Westminster Abbey
C. All awarded the Nobel Prize

11. Which element is named after an Austrian physicist who has the following epitaph written on her gravestone in Cambridge:

A physicist who never lost her humanity.

12. Which radioactive element was named after a husband and wife, the first ever married couple to win the Nobel Prize, and who were the husband and wife?

13. Below you will find clues to pairs of scientists linked simply by the fact that the second scientist's surname may be found as a string in the first (such as David Bohm and Georg Ohm):

A. Russian chemist of Periodic Table fame and Friar founder of genetics
B. Double Nobel Laureate (in Chemistry and Peace) and another Nobel Prize winner (in Physics) known for his Exclusion Principle
C. First woman to receive a medical degree in the USA and the Scottish physicist and chemist who discovered magnesium
D. English theologian, geologist and palaeontologist (who ate his way through much of the animal kingdom) and American biologist who shared the 2004 Nobel Prize for Physiology or Medicine for her work on the olfactory system
E. Indian mathematician who studied at Cambridge and was elected Fellow of the Royal Society (and is portrayed by Dev Patel in the 2015 film *The man who knew infinity*) and Indian physicist known for his work on light scattering (and a type of spectroscopy bears his name)
F. The Swedish chemist who discovered cobalt and the German alchemist who discovered phosphorus
G. Italian physicist who discovered Scorpius X-1, the first extra solar source of X-rays, and an Arctic and Antarctic explorer
H. German physician and biochemist who shared the 1922 Nobel Prize in Physiology and Medicine, and a German chemist who discovered the sulfur ring compound *thiophene*:

14. An elemental Wordsearch. Look for the listed scientists (or natural philosophers as they were known previously), up, down, diagonally or reverse. When you have completed this puzzle, you will be left with 10 cells that, when read left to right and down, will leave one chemist's way of arranging the elements (2 words). As an additional challenge, see how many scientists you can find without looking at the list below.

S	Ne	F	F	Ga	Ar	H	C	N	I	W	Er	Ca
O	W	La	V	O	I	S	I	Er	In	Es	Th	K
Re	Ar	Er	Y	W	Th	Ga	N	K	At	Er	Ni	C
N	B	Re	N	O	In	Li	Er	Cr	I	La	Co	I
S	U	N	Pr	Er	K	P	Po	Ne	N	Ds	La	Cr
O	Rg	La	F	N	He	P	Al	B	Se	Er	U	S
Re	K	N	Ra	O	I	I	O	Ra	N	I	S	I
N	Be	F	C	H	Ce	Rn	Se	He	U	V	Co	C
Se	Nd	La	U	Ra	B	As	Si	N	B	Cu	P	N
N	At	Ta	I	Cl	Au	Si	U	S	Be	Es	Er	Ra
Ag	Ne	S	Ar	B	Er	Y	Re	O	Fl	Rg	Ni	F
Ta	I	Ar	Ne	T	I	Se	Li	U	S	O	Cu	V
N	I	K	O	La	Te	S	La	Mo	Nd	Ge	S	Es

- Agnes Arber
- Arne Tiselius
- Ben Feringa
- Catherine Alice Raisin
- Erwin Chargaff
- Francis Crick
- Georges Cuvier
- Laura Bassi
- Nicolaus Copernicus
- Nikola Tesla
- Soren Sorensen
- Werner Heisenberg

And the following:

- Born
- Brahe
- Bunsen
- Clausius
- Florey
- Franklin
- Hippocrates
- Klaproth
- Land
- Lavoisier
- Mond
- Natta
- Perkin
- Urey
- Warburg

15. Which Oxford chemistry student studied X-ray crystallography under the Nobel Prize winner Dorothy Hodgkin and later became UK Prime Minister?

16. Who connects these 12 elements: americium, berkelium, californium, curium, dubnium, einsteinium, fermium, lawrencium, mendelevium, nobelium, rutherfordium, seaborgium?

A. Albert Ghiorso
B. Ernest O Lawrence
C. Glenn T Seaborg

17. What achievement links these scientists: John Bardeen, Marie Curie, Linus Pauling, Frederick Sanger, K Barry Sharpless?

18. The English physicist Robert Hooke (1635–1703) wrote the Latin anagram CEIIOSSOTTUU, unravelling as UT TENSIO, SIC UIS (as is the extension, so is the force) in one of his works in 1679 to establish priority in his discovery of what we now know as *Hooke's Law.*

Solve these anagrams to find the female scientists or inventors:

A. Doubter reeling (8, 1, 5)
American biochemist and pharmacologist who shared the 1988 Nobel Prize in Physiology and Medicine for innovative methods in drug design.

B. Emulate Chile diet (6, 2, 8)
French natural philosopher and mathematician who translated Newton's work and also published her own in 1740: *Foundations of Physics.*

C. Hardy realm (4, 6)
Hollywood actress who co-invented a radio guidance system for Allied torpedoes that used spread spectrum and frequency hopping technology to defeat the threat of jamming. She was subsequently inducted into the National Inventors Hall of Fame in 2014.

D. Ninja foundered (8, 6)
American biochemist who shared the 2020 Nobel Prize in Chemistry with Emmanuelle Charpentier 'for the development of a method for genome editing.'

19. Emilio Segrè (1905–89) was the first person to artificially create an element. After he scattered a small sample over his father's tomb, he wrote:

The radioactivity was miniscule, but its half-life of hundreds of thousands of years will last longer than any other monument I could offer.

What is the element?

A. Roentgenium
B. Seaborgium
C. Technetium

20. The following scientists met unfortunate ends. See if you can link the scientists with their premature endings:

1. Antoine Lavoisier, Chemist	**A.** Burned at the stake
2. Archimedes, Mathematician and Physicist	**B.** Cholera
3. Giordano Bruno, Philosopher	**C.** Fell under a horse-drawn cart
4. Marie Curie, Radiochemist	**D.** Guillotined
5. Nicolas Léonard Said Carnot, Physicist	**E.** Killed in battle
6. Pierre Curie, Physicist	**F.** Leukaemia
7. Thomas Midgley, Engineer	**G.** Strangled by his own invention
8. William Whewell, Polymath	**H.** Thrown from a horse

21. Which synthetic element first produced in the USA was nonetheless named after a Russian chemist during the Cold War?

22. A famous physicist in 1939 wrote a letter to the US President F. D. Roosevelt about the possibility of an atomic bomb, yet somewhat ironically had an element named after him which was first identified in 1952 in the fallout from the first successful test of a hydrogen bomb. What was the element?

23. Similar to a word ladder, each answer below provides letters that must appear in the answer immediately following it. The puzzle's final answer is a female French natural philosopher.

A.	_ _ _	Help (3)
B.	_ _ _ _	Notion (4)
C.	_ _ _ _ _	Perfect (5)
D.	_ _ _ _ _ _	Specify (6)
E.	_ _ _ _ _ _ _ _ _	Expound (9)
F.	_ _ _ _ _ _ _ _ _ _ _ _ _ _ _ _	French natural philosopher (6, 2, 8)

24. Chaim Weizmann (1874–1952) was a biochemist at the University of Manchester when he was asked in 1915 by Winston Churchill to help with the war effort with his fermentation process for the production of acetone from maize (used in the manufacture of the explosive cordite). Churchill is famously credited with the phrase '*Scientists should be on tap but not on top*' yet Weizmann did subsequently make it to the top. How and where?

25. Spiralling science and scientists

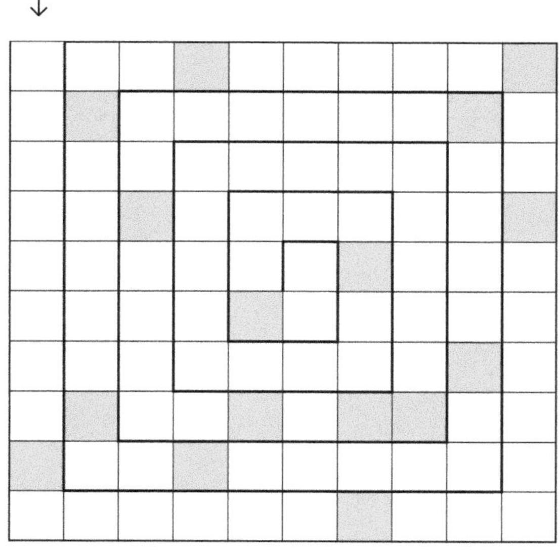

Starting at the cell immediately below the arrow and following the spiral around right to the centre in an anticlockwise direction, solve the clues in the order they appear. Apart from the first answer, every ensuing answer starts with the same two letters that the previous answer ends with. In addition, each cell can have 1 or 2 letters provided they are elemental symbols. Once complete, rearrange the elemental symbols in the yellow squares to reveal the name of an astronomer who had an element named after him.

- Italian mathematician and first woman to write a maths textbook (6)
- SiO_2 (6)
- Key element of organic chemistry (6)
- Physical chemist awarded the 1968 Nobel Prize in Chemistry for his work on thermodynamics (7)
- _ _ _ _ _ Schrödinger, quantum physicist (5)
- Hormone (7)
- Nickel–iron alloy with a low coefficient of thermal expansion (5)
- Swedish chemist and inventor of electrophoresis (4, 8)
- Acid found in lichens (5)
- Insect (9, 4)
- Zinc sulfide mineral (10)
- Synthetic element (10)
- Synthetic rubber (8)
- Geologic period (7)
- Interstellar cloud (6)
- Italian physicist and first woman to have a doctorate in science (5, 5)
- Elastomer (8, 6)
- Fatty acid found in rapeseed (6)
- Fossilised footprint (7)
- Nikola _ _ _ _ _ _, inventor (5)
- French polymath and author of *Celestial Mechanics* (7)
- Another name for hexa-1-decene (6)
- Precursor to the periodic table (8, 7)
- Garnet mineral (8)

26. In the puzzle below, replace the 3 question marks with letters to form two discoverers or co-discoverers of element(s) in two of the complete circles, plus the name of an element named after a scientist (who predicted another element) in the other complete circle. The four-letter word in the intersections is the name of that scientist.

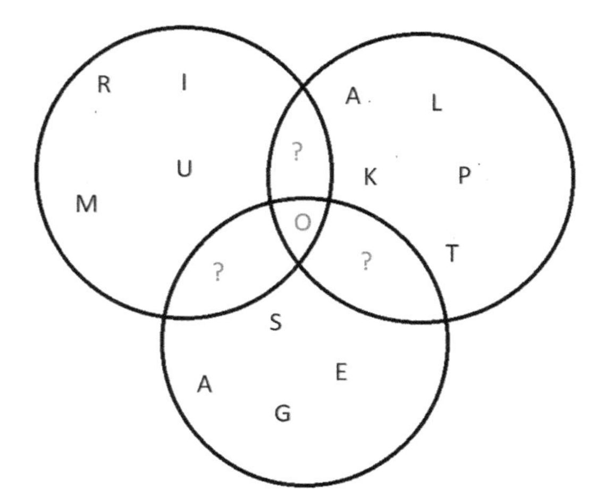

27. The following pairs of people (scientists or other famous people) are linked by the fact that the first person's surname is the second person's first name. Fill in the name shared:

A. Nikolaus August	_ _ _ _	(4) Hahn
B. Alfred Russel	_ _ _ _ _ _ _	(7) Carothers
C. Archer	_ _ _ _ _ _	(6) Heinrich Klaproth
D. Niels	_ _ _ _	(4) Tasman
E. Frances	_ _ _ _ _ _	(6) Schwarzenegger
F. Benjamin	_ _ _ _ _ _ _ _	(8) D Roosevelt
G. Susan	_ _ _ _ _ _ _	(7) Asch
H. Ruth	_ _ _ _ _ _ _ _	(8) Cumberbatch

28. Link the following scientists with their inventions (or co-inventions):

1. Sir Humphry Davy	**A.** Barometer
2. Sir James Dewar	**B.** Dynamo
3. Albert Einstein	**C.** Electric Battery
4. Michael Faraday	**D.** Kevlar®
5. Benjamin Franklin	**E.** Lightning Rod
6. Stephanie Kwolek	**F.** Miner's Safety Lamp
7. Evangelista Torricelli	**G.** Refrigerator
8. Alessandro Volta	**H.** Vacuum Flask

29. Acrostic. Solve the elemental clues across in the left-hand grid and reading down in the first column you will find a revolutionary chemist (two words). Transfer the letters in the left-hand grid into the right-hand grid as specified, and you will find a quotation about the chemist and the name of the mathematician who wrote it.

A Favourite of Victorian poisoners	**B.** The N in EPNS
C. Named after the father of Niobe	**D.** Named by the chemist in this puzzle
E. Halogen	**F.** Noble gas
G. Named after a village in Sweden	**H.** Song by Nirvana
I. Rare halogen	**J.** Named after a Scandinavian goddess
K. Smelly	**L.** Alloyed with previous element for pen nibs
M. Brimstone	**N.** Press
O. Named after a continent	**P.** Radioactive noble gas

	1	2	3	4	5	6	7	8
A								
B								
C								
D								
E								
F								
G								
H								
I								
J								
K								
L								
M								
N								
O								
P								

D1	A5	C6	D3		I1		
C8	N3	K3	O1	B1	I5		
C4	P4		A7	H6	I3		
E2	M4	M4		C1	H4	A1	C1
H4	D5	C2	P3		I4	I7	L4
C5							
H4	G5	J3	J5	L2	G1	E3	
D3	A4	P2	G2	A3			
L7	J2	D3		D6	O4	H3	
D4	E1	J1	I8		M2	M1	
J4	P5	K1	H3	H4	B5	M6	
M3	L1	B4	E6		N1	C1	
H1	P2	D4	P1	A1	F1	D4	F2

29. Constant reminders

Link the following constants with the scientists they are named after. The constant in each equation is the one in ***bold italics:***

1. $E = \boldsymbol{h}\nu$
2. $F = \boldsymbol{k_e}(Qq/r^2)\hat{e}_r$
3. $\boldsymbol{R} = m_e e^4/8\varepsilon_0^{\,2}h^3c$
4. $S = \boldsymbol{k} \log W$
5. $V = \boldsymbol{H_0}D$

A. Charles-Augustin de Coulomb (1736–1806)
B. Edwin Powell Hubble (1889–1953)
C. Johannes Rydberg (1854–1919)
D. Ludwig Boltzmann (1844–1906)
E. Max Planck (1858–1947)

Image courtesy of mycomicshop.com

In 1945, in the aftermath of the Second World War, scientists (and in particular those who had worked on the Manhattan Project) were hailed as heroes. Life magazine said '*They are men who wore the tunic of Superman and stand in the spotlight of a thousand Suns.*'

The Elementals was a Marvel Comics creation, comprising four immortal humanoids: Hydron, Lord of the Waters, Hellfire, wielder of flame, Magnum, master of the Earth, and Zephyr, mistress of the winds. Chemistry often gets a bad reputation, but this time Marvel takes it a few steps further. *Chemistro* was not just one villain, but was the alter ego of three villains (Curtis Carr, Archibald Morton and Calvin Carr, all of whom were involved with the *alchemy gun,* capable of transmuting matter from one form to another).

1. Which noble gas element is fictionally linked to Superman (and his dog!)?

2. *Superconductivity* is a phenomenon where electrical resistance of a solid disappears and magnetic flux fields are expelled below a certain temperature. To date, this has led to the award of five Nobel Prizes in Physics. The Dutch physicist Heike Kamerlingh Onnes (1853–1926) was the first of these Nobel Laureates (awarded in 1913 after his 1911 discovery), having been the first to discover superconductivity.

An isotope of this gas also led to the discovery of *superfluidity* in 1937 by Soviet physicist Pyotr Kapitsa (1894–1984), who also won the Nobel Prize in Physics (in 1978). Superfluidity describes a fluid with zero viscosity, so it flows with no loss of kinetic energy. What was the gas?

A. Deuterium
B. Helium
C. Hydrogen

3. Marvellous!

   Link these Marvel Comics superheroes and villains (with elemental connections) with their alter egos and some of their superpowers:

   1. Arsenic
   2. Cobalt Man
   3. Iron Man
   4. Quicksilver
   5. The Silver Surfer

   A. Norrin Rad (who wields the *Power Cosmic* and can travel through time)
   B. Gertrude Yorkes (who has a telepathic bond with her dinosaur *Old Lace*)
   C. Pietro Maximoff (who can run at supersonic speeds)
   D. Tony Stark (who has a mechanised suit of armour)
   E. Ralph Roberts (who achieved superhuman size and power after being bombarded with radiation)

4. *Supercooling* involves lowering the temperature of a liquid or gas below its freezing point without becoming a solid. Which well-known drinks company briefly had special vending machines with supercooled drinks which turned to slush on opening?

5. Which element has the lowest boiling point of any substance, and is used to supercool the Large Hadron Collider, MRI scanners and to cool the hydrogen used in some rockets?

   A. Helium
   B. Hydrogen
   C. Neon

6. Even more Marvellous!

   Link these Marvel Comics' superheroes and villains (with scientific or pseudo-scientific connections) with their alter egos, and some of their superpowers:

   1. Alchemy
   2. Asbestos Man
   3. Electro
   4. Graviton
   5. Microbe
   6. Molecule Man
   7. Nitro

   A. Robert Hunter (who can transform his body into a gaseous state, explode and reconstitute himself)
   B. Franklin Hall (who had a PhD in Physics and could manipulate gravity)
   C. Maxwell Dillon (one of Spider-man's enemies who turned villain after being struck by lightning when working on a power line, making him a human electric capacitor with devastating results)
   D. Thomas 'Jellybean' Jones (who can reduce the chemical composition of anything he touches into its elemental components)
   E. Dr Orson Karloff (with a superhuman ability to analyse chemicals, holds a PhD in analytical chemistry, and created a flame-resistant armour)
   F. Zachary Smith Jr (who is a mutant with the ability to communicate with germs and other microscopic organisms)
   G. Owen Reece (who had the ability to mentally manipulate molecules)

7. Which phenomenon is also known as *boiling retardation*?

   A. Supercooling
   B. Superfluidity
   C. Superheating

8. *Superheavy* was a short-lived *supergroup* comprising Mick Jagger, Joss Stone, Dave Stewart, A. R. Rahman, and Damian Marley. *Superheavy elements* are usually short-lived too. Most scientists use the term *superheavy element* to describe an element with more than 100 protons in its nucleus.

   Name these superheavy elements from their anagrams, but to make it slightly harder, each anagram has one letter too many. Once you've solved the anagrams, rearrange these four additional letters to spell the name of another element (which is not superheavy).

   A. Bogus mailer
   B. Nonsense tie
   C. Rumoured fright
   D. Sane godson

9. Similar to a word ladder, each answer below provides letters that must appear in the answer immediately following it. The puzzle's final answer is Superman's weakness.

A. _ _ _	Interfere (3)
B. _ _ _ _	Target (4)
C. _ _ _ _ _	Of poor quality (5)
D. _ _ _ _ _ _	Verse (6)
E. _ _ _ _ _ _ _	S, or the degree of disorder in a system (7)
F. _ _ _ _ _ _ _ _ _ _	Superman's weakness (10)

P	Ta	Er
V	Se	Ca
Ni	U	In

10. Using all 9 elemental symbols in the grid above (and *not* separating the letters of a two-letter symbol), spell out the name of a superhero (two words) whose persona merged with several hosts and was the guardian and protector of Eternity.

Then answer these questions using some of the symbols always including the middle symbol and not switching letters with 2-letter symbols (using each symbol once and once only in each word), for instance [Ca]V[Er] would be allowed but [Ca]V[eS] would not.

A. Nose bones (5)
B. Cut (5)
C. Poem (5)
D. Examine (6)
E. Milk protein (6)
F. Opposite (7)

11. Elon Musk's SpaceX's *Super Heavy* booster powers the Starship rocket. Its Raptor engines use a fuel mix of liquid oxygen and a liquid hydrocarbon. What is that hydrocarbon?

A. Methane
B. Ethane
C. Propane

12. *Supersymmetry*† (or SUSY) predicts a partner particle for each particle in the Standard Model of Quantum Mechanics. For instance an electron will have a partner *selectron* (a boson) and the photon will have a partner *photino* (a fermion). What are these supersymmetric partners called?

A. Sparticles
B. Mirror particles
C. Reflectives

† *Supersymmetry* is also a song by Arcade Fire.

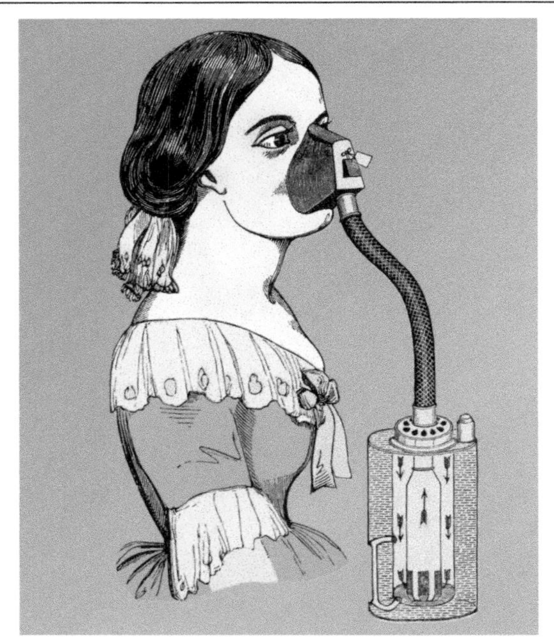

John Snow's Chloroform Inhaler, 1850s. © Science Source/Science Photo Library

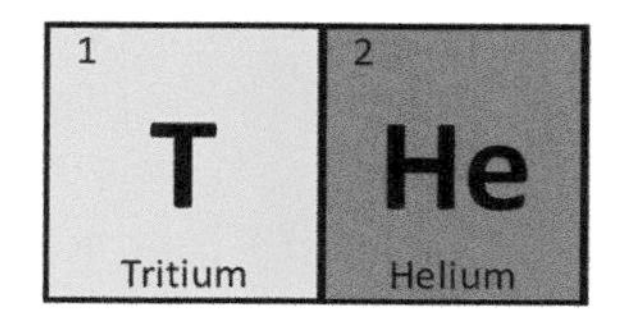

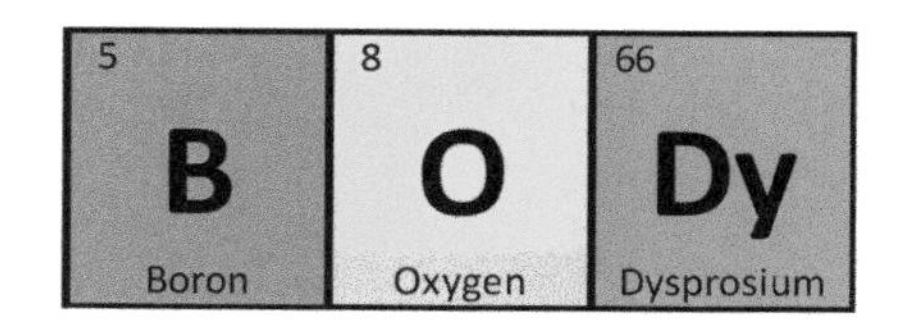

From quicksilver cures to other quack remedies, the history of medicine makes for gripping and gruesome reading. Yet despite the often fatal trials and errors, as the science progressed, many of us owe our lives and our quality of life to modern anaesthetics, antiseptics, antibiotics, analgesics and other assorted pharmaceuticals (and of course better diets), and the endeavour and courage of those physicians (and their patients) who went before us.

1. Which element is at the centre of the haemoglobin molecule which carries oxygen in the blood?

2. Which halogen is essential for functioning of the thyroid? The deficiency disease can lead to goitre (a swelling on the neck), once called *Derbyshire Neck* due its prevalence in that area of England.

3. Many years ago 'corrosive sublimate' (recommended by none other than Robert Boyle) was used in the treatment of venereal diseases. Indeed, in such times, a popular saying went 'a night with Venus leads to a life with (the element)'. And like Venus, this element does have a heavenly body. What is the element?

4. Vitamin B_9 (above) is a dietary supplement often taken by expectant mothers to aid a healthy pregnancy. In the USA and other countries, it is also added to staples such as cereals, rice, pasta and bread. What is it more normally known as?

5. What compound which builds up in the body during high-intensity exercise, causing the muscles not to work so effectively, is also found in fermented milk products?

Scientific Researches! New Discoveries in Pneumaticks! An Experimental Lecture on the Powers of Air, 1802. James Gillray

6. In 1799, the Cornish chemist Sir Humphry Davy (1778–1829) held a party of the great and the good to try out the effects of a new gas he'd recently produced. This gas has since been used for pain relief (such as in toothache, which Davy himself used). One of the celebrities at the party, the English poet Samuel Taylor Coleridge (1772–1834) subsequently recounted the effects of his inhalation of the gas:

   The first time I inspired.....I felt an highly pleasurable sensation of warmth over my whole frame, resembling that which I remember once to have experienced after returning from a walk in the snow into a warm room. The only motion which I felt inclined to make, was that of laughing at those who were looking at me. My eyes felt distended, and towards the last, my heart beat as if it were leaping up and down...I fixed my eyes on some trees in the distance, but I did not find any effect except that they became dimmer and dimmer, and looked at last as if I had seen them through tears....I could not avoid, nor indeed felt any wish to avoid, beating the ground with my feet; and after the mouth-piece was removed, I remained for a few seconds motionless, in great extacy (sic).

   What is the gas?

7. Which toxic element adds insult to injury when alloyed with lead to make lead shot harder than it would otherwise be, but was also used to form the 'magic bullet' to cure syphilis in the early 20th century?

8. Link these vitamins and essential elements with their associated deficiency diseases:

1. Anaemia	A. Vitamin A
2. Beriberi	B. Vitamin B_1
3. Blindness	C. Vitamin B_3
4. Goitre	D. Vitamin C
5. Osteoporosis	E. Vitamin D
6. Pellagra	F. Calcium
7. Rickets	G. Iodine
8. Scurvy	H. Iron

9. Amino acids are so named since they all contain a central carbon atom bonded to an amine group and an acid group. There is a third variable group attached to the central carbon called the side chain, and amino acids are the building blocks of proteins. There are 20 amino acids that the human body needs to function properly. Work out some of them from their anagrams below:

A. Emulating
B. Use nice oil
C. No peril
D. Spare again
E. Nice stye
F. Hit nominee

10. Of the 20 amino acids, 9 of them are classed as *essential* since the human body cannot synthesise them fast enough to match demand, so they need to come from the diet. Three of these may be made from the whole circles below once you have filled in the remaining intersections, and the 4-letter word in the middle intersections is an essential element. What are the 3 essential amino acids and the essential element?

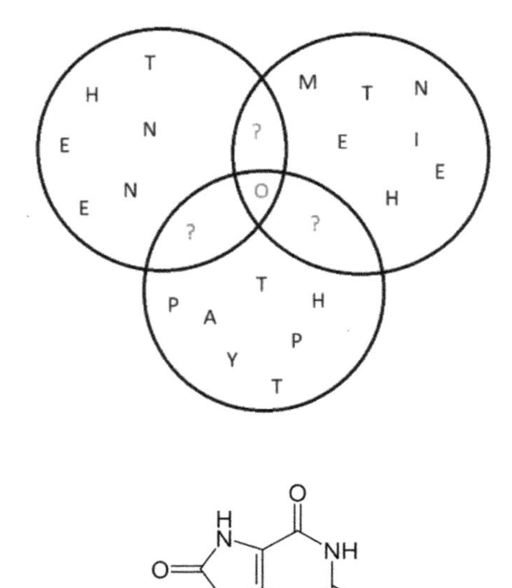

11. The above compound can be implicated in diabetes, gout and kidney stones. What is it?

12. Trihalomethane, $CHCl_3$ was once used as an anaesthetic, and indeed Queen Victoria was administered this during childbirth. What is the compound's common name?

13. Although toxic, this element was used medicinally, particularly by the English quack doctor Joshua Ward. Ward began to sell his medicines in London in 1733, and the following saying relates to Ward's Pills and Drops: '*Before you take his Drop or Pill, Take leave of your friends and make your will.*' What is the element?

14. *'I remember reading how the town authorities in the little town of Carlisle rendered their sewage sweet smelling, by treating with carbolic acid. And I wondered. Patients with compound fractures, fractures where the skin of the fracture site is broken nearly always died of sepsis. I had an idea, an hypothesis – perhaps infected fractures were like sewage.'*

Lord Joseph Lister speaking to Winnipeg Medical Students, 1898

Lister was the pioneer of antiseptic surgery (using carbolic acid). What is carbolic acid now more commonly known as?

15. Which element's name comes from the Greek for 'masculine' or 'virile' and was recommended for lovemaking in the *Kama Sutra of Vatsyana* (a Sanskrit love manual of about 3rd century A.D., translated by Sir Richard Burton and F. F. Arbuthnot)?

16. *Itai-itai disease* is named after the Japanese for *ouch ouch*, due to the joint pain experienced. There was an outbreak of this disease in the 1960s, in the Jinzu river basin, due to contamination of rice crops from a local zinc mine. What is the element (not zinc) that was the causative agent for this disease?

17. Which English chemist and meteorologist instructed his assistant to remove his eyeballs after his death to investigate the causes of his colour-blindness (and colour-blindness is sometimes referred to by his name)?

18. Which famous physicist's brain and eyes were removed after death, the brain being dissected into 240 pieces for subsequent research?

19. 99% of the human body is comprised of just six elements. What are they?

20. About 0.85% of the remainder is comprised of just five additional elements. What are they?†

21. *Argyria* is a condition that can turn your skin a blue-grey colour, and indeed did do with the US politician Stan Jones (b 1943), who dosed himself up with a certain element in colloidal form since he considered it to be an antibiotic. What was the element?

---

†Both questions 19 and 20 are about the human body before the addition of breast enhancements, hip replacements, pacemakers *etc.*!

UK Postage Stamps celebrating the science fiction and prescience of H G Wells

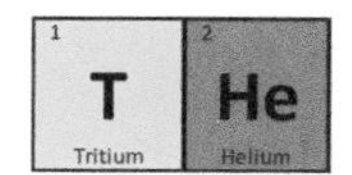

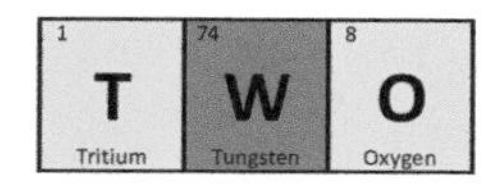

Cultures

The Two Cultures? Sounds like something growing on a Petri Dish?† No, it's referring to a famous lecture and essay (full title *The Two Cultures and the Scientific Revolution*) in 1959 by British novelist and scientist C P Snow (1905–1980) whose thoughts are made abundantly clear in this excerpt:

> *A good many times I have been present at gatherings of people who, by the standards of the traditional culture, are thought highly educated and who have with considerable gusto been expressing their incredulity at the illiteracy of scientists. Once or twice I have been provoked and have asked the company how many of them could describe the Second Law of Thermodynamics. The response was cold: it was also negative. Yet I was asking something which is the scientific equivalent of: Have you read a work of Shakespeare's? I now believe that if I had asked an even simpler question – such as, What do you mean by mass, or acceleration, which is the scientific equivalent of saying, Can you read? – not more than one in ten of the highly educated would have felt that I was speaking the same language. So the great edifice of modern physics goes up, and the majority of the cleverest people in the western world have about as much insight into it as their neolithic ancestors would have had.*

Many schoolchildren (at least in the UK) have had to make the difficult choice on whether to choose the Arts or the Sciences during schooling, and on which route to take after that. Yet back in the early days of the 19th century, the arts and sciences intermingled, with the Romantic Poets rubbing shoulders with the scientists of the day.

For instance the poet Samuel Taylor Coleridge (1772–1834), wrote:

> *I shall attack Chemistry like a Shark.*

on his plans to set up a chemistry laboratory in the Lake District with poet William Wordsworth (1770–1850), and chemist and inventor Sir Humphry Davy (1778–1829). He never did, or at least we don't know he did (and it's 200 years since).

This section deals with those instances where the arts and sciences have embraced each other, and of course there are some references to sci-fi, perhaps the ultimate combination of scientific fact and artistic fiction (although some sci-fi stories have been amazingly accurate in their predictions).

---

† The Petri Dish was named after a scientist, like many pieces of lab equipment. The German bacteriologist Julius Richard Petri (1852–1921) is credited with inventing it while working as an assistant to the bacteriologist Robert Koch (1843–1910). Koch was awarded the Nobel 1905 Nobel Prize in Physiology and Medicine for his work on tuberculosis.

1. Having exhausted the humdrum silver, gold and platinum as metals worthy of a record of his achievements, Paul McCartney was awarded a disc made of a far pricier metal on October 24th 1979 for being the world's most successful song writer. What was the metal?

   A. Palladium
   B. Osmium
   C. Rhodium

2. Link the scientifically connected song or album titles with the singer or group:

1. Astronomy Domine	**A.** Blackalicious
2. Atomic	**B.** Blondie
3. Bunsen Burner	**C.** Coldplay
4. Chemical Calisthenics	**D.** John Otway
5. Chemistry	**E.** Sting
6. The Scientist	**F.** Paul McCartney and Wings
7. Magneto and Titanium Man	**G.** Pink Floyd
8. Mercury Falling	**H.** Semisonic
9. Neon Wilderness	**I.** The Pointer Sisters
10. Neutron Dance	**J.** The Verve
11. She Blinded Me With Science	**K.** Thomas Dolby

3 Fill in the element blanks in these film titles:

A. _ _ _ _ _ _ _ _ Blonde (1931)
B. _ _ _ _ _ _ _ and Old Lace (1943)
C. _ _ _ _ _ _ Copy (1981)
D. _ _ _ _ _ _ _ _ _ _ Baby (1987)
E. _ _ _ _ _ _ _ Rising (1998)
F. The _ _ _ _ _ _ _ Kid (2004)
G. My _ _ _ _ Bed (2008)
H. _ _ _ _ Man 3 (2013)
I. The _ _ _ _ Spectrum (2017)

4 What precious metallic element is the French-English writer Hilaire Belloc (1870–1953) referring to in his ditty *The Hippopotamus*:

I shoot the Hippopotamus
With bullets made of - - - - - - - -,
Because if I use leaden ones
His hide is sure to flatten 'em.

5 Elements rock!

Fill in the missing words from the world of rock'n'roll.

A. ______back
B. ____ Belly
C. ______ Bullet Band
D. ____frapp
E. Freddie _______

6 Link these facts with the science fiction writer:

1. Isaac Asimov (1920–1992)
2. Ray Bradbury (1920–2012)
3. Arthur C. Clarke (1917–2008)
4. Joanna Russ (1937–2011)
5. Jules Verne (1828–1905)
6. H. G. Wells (1866–1946)

The facts:

a. One of this undersea explorer's science fiction books was used by the rocket scientist, Werner von Braun, to convince President John F. Kennedy of the possibility of going to the moon.
b. A writer of non-fiction and fiction (and fiction of many genres, not just science fiction), nominated for the Nobel Prize in Literature four times. A diabetic, he co-founded The Diabetic Association (now *Diabetes UK*). He was also taught by English biologist Thomas Henry Huxley (1825–1895) at the Royal College of Science.
c. French novelist, playwright and poet. Also a diabetic, shot by a nephew in the leg which resulted in a permanent limp.
d. American writer and screenwriter whose first pay as a writer at age 14 was for a joke written for the American comedian Georg Burns.
e. Born in the Bronx, studied with lepidopterist and writer Vladimir Nabokov.
f. A professor of biochemistry at Boston University, wrote many popular science books as well as science fiction.

7 In the TV Series *Breaking Bad* the main character Walter White, a chemistry teacher turned drug maker, disposes of some bodies in a bathtub with the help of Jesse using what?

A. Hydrochloric acid
B. Hydrofluoric acid
C. Sulfuric acid

8 The 2021 Movie *Minamata,* starring Johnny Depp, is about a major pollution incident in Japan. What toxic metal was involved?

$$\mathrm{RCOO^-\ Ag^+} \xrightarrow[\mathrm{CCl_4}]{\mathrm{Br_2}} \mathrm{R{-}Br}$$

9 The above reaction is named after a Russian chemist Alexander Borodin (1833–1887) who is better known for something other than chemistry. What is it?

10 Link these elements of fiction with the clues:

1. Martin Amis invented this 'element' to describe the discomfort of his extensive dental work in his memoir *Experience* (Hyperion 2000):

   'A new shoe: in your mouth. No, a football boot: in your mouth. A boot forged from an element quite new to the periodic table: an element called _ _ _ _ _ _ _ _ .'

2. An element named after a country in H. G. Wells' novel *Tono Bungay* (1909), where George Ponderovo is apprenticed to his Uncle Edward, a chemist who invents a bogus medicine and earns a vast fortune.
3. A radioactive element in the 1936 science fiction horror film *The Invisible Ray.* Where the scientist Dr Janos Rukh (played by Boris Karloff as *The Luminous Man)* glows in the dark (and his touch becomes lethal) when he is exposed to the element in a meteorite.
4. An element invented by Jules Verne in his early novel *The Chase of the Golden Meteor* (published posthumously in 1908).
5. The hydrogen isotope ^{4}H, a fissionable material used in a *Q-Bomb* in Leonard Wibberley's 1955 novel *The Mouse that Roared.* (Note that this highly unstable isotope has been synthesised in the laboratory by bombarding tritium with deuterium).
6. The element in the black comedy *Dr Strangelove* which is used in the Russians' doomsday device.
7. Element created by Disney for a 1957 episode of *Uncle Scrooge*, said to be the world's rarest element and which tastes different every time one tries it.
8. The highly radioactive element in Clive Cussler's *Raise the Titanic!* (which was later made into a film).

A. Bombastium
B. Byzanium
C. Canadium
D. Cobalt Thorium G
E. Nausium
F. Quadium
G. Radium X
H. Xirdalium

11 Who connects these other fictional elements: *bolognium*, *frinkonium* and *jumbonium*?

12 English playwright and poet William Shakespeare died in 1616. Only a small percentage (about 12%) of the 118 elements in the current Periodic Table were known to him (the first element discovered after his passing being phosphorus in 1669). One of the elements known in Shakespeare's time contains the name of one of his characters. What is the element and who is the character?

13 Which English classical composer, keen crossword puzzler, and amateur chemist invented and patented an apparatus (pictured above) for the production of hydrogen sulfide, H_2S (that smells like rotten eggs at low concentrations)?

14 The American science fiction and fantasy author Ursula Le Guin (1929–2018) wrote '*First sentences are doors to worlds*' in her essay *The Fisherwoman's Daughter.* Match these opening lines to the science fiction or science-related novels and their authors:

1. A screaming comes across the sky.
2. It was a bright cold day in April, and the clocks were striking thirteen.
3. It was a pleasure to burn.
4. The drought had lasted for ten million years, and the reign of the terrible lizards had long since ended.
5. The story so far; In the beginning the Universe was created.
6. The year 1866 was signalised by a remarkable incident, a mysterious and puzzling phenomenon, which doubtless no one has yet forgotten.
7. This is a tale of a meeting of two lonesome, skinny, fairly old white men on a planet which was dying fast.
8. When a day that you happen to know is Wednesday starts off by sounding like Sunday, there is something seriously wrong somewhere.
9. You will rejoice to hear that no disaster has accompanied the commencement of an enterprise which you have regarded with such evil forebodings.

A. *1984* by George Orwell (1903–1950)
B. *2001 A Space Odyssey* by Arthur C. Clarke (1946–2008)
C. *Breakfast of Champions* by Kurt Vonnegut (1992–2007)
D. *Fahrenheit 451* by Ray Bradbury (1920–2012)
E. *Frankenstein* by Mary Shelley (1797–1851)
F. *Gravity's Rainbow* by Thomas Pynchon (b 1937)
G. *The Day of the Triffids* by John Wyndham (1903–1969)
H. *The Restaurant at the End of the Universe* by Douglas Adams (1952–2001)
I. *Twenty thousand leagues under the Sea* by Jules Verne (1828–1905)

15 Which diatomic alkali metal actually exists (in a gaseous state) but was also the name of an invented substance that powered Star Trek's *USS Enterprise*?

16 The Nobel Prize in Literature has yet to be awarded to a writer whose main genre is science fiction. However, some have dabbled. Link these winners of the prize (with year of award) with their sci-fi or science-infused works:

1. Kazuo Ishiguro (2017)	**A.** Arrowsmith
2. Doris Lessing (2007)	**B.** The Making of the Representative for Planet 8
3. Bertrand Russell (1950)	**C.** With the Night Mail: A Story of 2000 A.D.
4. Sinclair Lewis (1930)	**D.** The ABC of Atoms
5. Rudyard Kipling (1907)	**E.** Klara and the Sun

17 Another winner of the Nobel Prize in Literature, William Golding was a friend of the scientist James Lovelock. Golding suggested a name from Greek mythology for Lovelock's hypothesis about the Earth being a self-regulating organism. What was the name?

18 Which element's symbol is the name given to Japan's classical theatre?

19 Link these poetic first lines to the science-inspired poems and their authors:

1. Because I love the very bones of you
2. If all a top physicist knows
3. If you look in old chemistry books
4. Radium is my element
5. Who affirms that crystals are alive?
6. You crack an atom, what's left? Particles,
7. Your coffee grows cold on the kitchen table
8. Your famous apple

A. *Calcium* by Deryn Rees-Jones
B. *Curie* by Kelly Cherry
C. *Dear Isaac Newton* by Charles Simic
D. *Entropy* by Neil Rollinson
E. *After Reading a Child's Guide to Modern Physics* by W H Auden
F. *Modes of Representation* by Roald Hoffman
G. *Snow* by John Davidson
H. *The Cloud Chamber* by Philip Gross

20 Which pop star reportedly slept in an oxygen chamber and died a few weeks before his sell-out tour at London's O_2 Arena in 2009?

21 The elements in Group 17 of the Periodic Table are called halogens (meaning 'salt producers') as they form salts (like common salt, sodium chloride). Which halogen did Leonard Cohen sing about in his album *Death of a Ladies' Man*?

22 Surnames are shared by scientists and other famous people. Link these surnames to the clues:

1. Inventor of the Cloud Chamber/British Prime Minister
2. Philosopher and alchemist/Figurative painter
3. Inventor of the cyclotron/Author of *Seven Pillars of Wisdom*
4. Discoverer of penicillin/Inventor of James Bond
5. Deviser of modern geological timescales/Fictional detective
6. Co-discoverer of DNA structure/English actress in the mini-series *Chernobyl*

A. Holmes
B. Lawrence
C. Watson
D. Wilson
E. Bacon
F. Fleming

23 The 1965 war film *The Heroes of Telemark* is based on the memoirs (*Skis against the Atom*) of the Norwegian resistance soldier Knut Haukelid. In it, the production of a critical component of a potential German atomic weapon was sabotaged. What was the component?

A. Uranium-235
B. Heavy Water, D_2O
C. Plutonium-239

24 Match the elements to these works of fiction:

1. The ? Drum by Günter Grass
2. The Man in the ? Mask by Alexandre Dumas
3. The ? Beech by Maeve Binchy
4. Altered ? by Richard Morgan
5. ? Daughters by Iain Sinclair

A. Carbon
B. Copper
C. Iron
D. Radon
E. Tin

25 William Morris (1834–1896) initiated the *Arts and Crafts* design movement in 1861. Amongst other things, he became renowned for his beautiful wallpapers, some of them exhibiting vivid greens. However, there was a potential catch. The green pigment, known as *Scheele's Green* (named after its inventor, the Swedish chemist Carl Wilhelm Scheele) was highly toxic due to the presence of a certain element. Amongst other things, it was blamed for the death of Queen Victoria's parrot and Emperor Napoleon, although these are matters of some conjecture. What was the element?

26 A sometimes controversial British artist fell out with the *Royal Pharmaceutical Society of Great Britain* over a Notting Hill, London restaurant called *Pharmacy*, opened in 1997. Calling the restaurant *Pharmacy* contravened the *Medicines Act 1968* so the name was changed to anagrams: *Army Chap* and then *Achy Ramp.* The society also claimed that the pill bottles and medical paraphernalia on display could confuse members of the public who were looking for a real pharmacy. The restaurant closed in 2003 but the artist reportedly earnt over eleven million pounds when the artworks and artefacts were sold. Who was the artist?

A. Tracey Emin
B. Lucian Freud
C. Damien Hirst

27 The above artist is also well known for animal artwork in glass panel display cases, such as *The Physical Impossibility of Death in the Mind of Someone Living* which uses a tiger shark. These installations are all preserved in a solution of which chemical?

A. Acetic acid
B. Carbolic acid (phenol)
C. Formaldehyde

28 After crystallising engines in his 2004 artwork *Nunhead*, in 2008 the British artist Roger Hiorns created a work called *Seizure* in an unoccupied council flat in London. 75 000 litres of a chemical solution were pumped into the waterproofed flat to create a blue crystalline growth on the walls, floors, ceiling and bath. This artwork was a source of inspiration for Icelandic singer-songwriter Björk in her song *Crystalline.* What was the chemical?

A. Cobalt aluminate
B. Copper sulfate
C. Prussian blue

29 *Nuclear Energy* (above) is a bronze sculpture on the campus of the University of Chicago, at the site of the world's first nuclear reactor, Chicago Pile-1. Who was the sculptor?

A. Marcel Duchamp
B. Barbara Hepworth
C. Henry Moore

30 Which UK Radio DJ has written a series of novels about a 14-year old boy called *Itchingham Lofte,* on a quest to find all of the elements of the Periodic Table?

A. Zoe Ball
B. Tony Blackburn
C. Simon Mayo

31 Which poet and Nobel Prize winner claimed in 1919 that '*poetry is a science*', likening the poetic process to catalysis as follows, '*the action which takes place when a bit of finely filiated platinum is introduced into a chamber containing oxygen and sulphur dioxide*' with the poet's mind being the platinum catalysing poetry out of pre-existing material.

A. W B Yeats
B. T S Eliot
C. Rudyard Kipling

32 Which artist wrote the following after it was shown that most of the atom was empty space:

'*The crumbling of the atom was to my soul like the crumbling of the whole world. Suddenly the heaviest walls toppled. Everything became uncertain, tottering and weak. I would not have been surprised if a stone had dissolved in the air and became invisible. Science seemed to be destroyed.*'

A. Wassily Kandinsky
B. Pablo Picasso
C. Paul Klee

Alchemy

1 LEAD, LOAD, GOAD, GOLD

(This puzzle is an example of a Word Ladder, a puzzle invented by the writer and mathematician Charles Lutwidge Dodgson (also know as Lewis Carroll) on Christmas Day 1877. He called them *Doublets* and they were first published in *Vanity Fair*. For more on Lewis Carroll, see the Particles chapter).

2 The alchemist was Paracelsus (c1493–1541). The answers are A Pelican, B Antimony (Sb), C Robert, D Alchemy, E Chariot, F Egypt, G Lead, H Saturn, I Urine, J Sulfur (S). The quotation is '*No man can transmute any matter if he is not transmuted himself.*'

3 Sir Isaac Newton (1642–1727).

4 Tycho Brahe (1546–1601) and the element is silver, Ag. Talented, eccentric and quarrelsome, Brahe lost most of his nose in a duel at the age of 19 when he fought over a mathematical dispute, and he made himself a replacement out of silver which can be seen in contemporary portraits of him. However, curious archaeologists found a green crust on the front of Brahe's skull when digging him up, suggesting a copper prosthetic. Perhaps he kept the silver one just for best occasions.

5 Gibberish!

6 1G, 2F, 3B, 4E, 5C, 6D, 7A

7 1F, 2C, 3E, 4G, 5A, 6D, 7B

8 B Dog's Mercury (*Mercurialis perennis*), which also acts as a diuretic when eaten by cattle, turning their urine a reddish-blue colour.

9 Fool's Gold. Its brass-yellow lustre gives the mineral a superficial resemblance to gold.

10 Bain-marie, translating as 'Mary's bath'

Around the House

1 Americium, Am (specifically americium-241) which decays to neptunium, Np (specifically neptunium-237), emitting alpha-particles. These alpha particles ionise molecules in the air which then carry a current between electrodes. Smoke particles will interrupt this current, thus activating the alarm. When the smoke detector is new, it is close to 100% americium. After 30 years, about 5% of the americium atoms will have decayed to neptunium.

In the 1990s. The boy scout David Hahn extracted americium from many smoke detectors to attempt to build a nuclear reactor in his shed. His other sources of radioactive material were thorium from gas mantles, radium from clocks and tritium from gunsights. (His story may be read in Ken Silverstein's 2004 book *The Radioactive Boy Scout*).

2 Carbon monoxide, CO which can come from some poorly maintained heating appliances. More people die from CO exposure than any other type of poisoning and if you don't have a CO monitor fitted already where you live, get one!

3 Graphite, an allotrope of carbon

4 1D, 2E, 3F, 4C, 5G, 6A, 7B

5 Drain cleaner (sulfuric acid) and bleach (sodium hypochlorite) as they can react to produce chlorine (Cl_2) which was used as a poison gas in the First World War.†

---

† (However, you can safely construct your own 'volcano' with a cone of sand with a hole made in the top, and filling the hole with baking soda, and then adding the vinegar (best done outside!) as the author's son did many years ago).

6 Acetone, ethylene glycol and sodium hypochlorite. Permitted food additives in the UK are allotted 'E numbers' which are listed in the ingredients. For a full UK list, search for 'Food Standards Agency.'

7 Radon, Rn

8 A Cerium, Ce. The cerium oxide acts as an oxidation catalyst.

9 B Tin, Sn

10 Argon, Ar

11 Radox

12 Water, H_2O. Water vapour in the air is sufficient to polymerise.

13 Cuprinol. From Latin *cuprum* for copper and *ol* for oil. The original preservative contained a copper compound and was known as Cuprinol Wood Preserver Green.

14 WD-40. The name stands for Water Displacement 40th Formula.

15 Bismuth[‡], which is an anagram of *this bum.* Fitting, since it is used in the treatment of haemorrhoids and perhaps gives another meaning to the term *atomic piles.*

For those interested, the chemical equation for the transformation is:

$Bi(OH)_2NO_3$ (pearl white) + $3H_2S = Bi_2S_3$ (black precipitate) + $2HNO_3 + 4H_2O$

[‡] This pearl-white had several aliases: magistery of bismuth, blanc d'Espagne, Spanish white, white cosmetic, and bismuth subnitrate.

Atmospheric

1 Nitrogen, N_2

2 Argon, Ar. Argon was discovered as a result of its being in our atmosphere in 1894 by Lord Rayleigh (1842–1919) and Sir William Ramsay (1852–1916). Both were awarded a Nobel Prize in 1904, in Physics and in Chemistry respectively.

3 1C, 2A, 3B

4 Venus

5 Oxygen, O (or O_2 and O_3 – ozone – in the atmosphere). The play is about the discovery of the element in the 18th century.

6 The Cloud Chamber (or Wilson's Cloud Chamber)

7 C Silver iodide, AgI. Cloud seeding in other parts of the world have since taken place both for peacetime and military objectives (such as *Project Stormfury* and *Operation Popeye* respectively). Silver iodide is highly photosensitive and was used by Louis Daguerre (1787–1851) in his early photographs called *Daguerrotypes.*

8 B John Dalton (1766–1844)

9 C Tottenham Hotspur

10 Chlorofluorocarbon

11 When fossil fuels are burnt, sulfur dioxide (SO_2) and nitrogen oxides (NO_x) are produced. These air pollutants then react with water, oxygen and other compounds to form sulfuric acid (H_2SO_4) and nitric acid (HNO_3). However, there are other (non-anthropogenic) sources of acid rain (such as volcanoes).

Beasts

1 The alternative (German) name for the element reading down is Wolfram (animals wolf and ram) which gives rise to the elemental symbol W. The animals are weasel, seahorse, flamingo, firecrest, frog, llama, marmoset.

2 The ant. Formic acid was first obtained in a fairly pure state by German chemist Andreas Sigismund Marggraf in 1749, by distilling red ants (*Formica rufa*) and redistilling the product. This acid occurs freely in ants, in the stings of bees and in the stinging nettle.[§]

3 Copper, Cu. Somewhat interestingly, the blood clam (*Tegillarca granosa*), also known as the blood cockle does contain haemoglobin and is considered a delicacy in Asia and Central America, eaten raw or briefly cooked. The haemoglobin allows the blood clams to survive in low-oxygen environments.

4 Cochineal. The cochineal is a scale insect and lives on cacti. The largest exporter of this dye is currently Peru. James Lovelock (who passed away on his 103rd birthday in 2022) was one of the author's science heroes (and he has been lucky enough to meet him). Lovelock describes his preparation of the dye as a young scientist by boiling one hundredweight of beetles with dilute acetic acid in his autobiography *Homage to Gaia* (2000, Oxford University Press).

Cochineal is also known as Ponceau 4R, or in the UK, food additive E124. (Ponceau is 17th century French for poppy-coloured and is the generic name for a family of azo dyes). It is one of the six permitted food colours which the UK Food Standards Agency (food.gov.uk) recommends manufacturers to find alternatives due to a possible link with hyperactivity in children, and if used in food and drink, to carry a warning on the packaging 'May have an adverse effect on activity and attention in children.' Also known as the 'Southampton six' due to a Southampton University study commissioned by the FSA (published in 2007), the other colours in this list are Sunset Yellow FCF (E110), Quinoline Yellow (E104), Carmoisine (E122), Allura Red (E129) and Tartrazine (E102).[¶]

[§] Formica is also the trade name of a laminated composite material invented by Westinghouse Electric Corporation in the USA in 1912, originally as a replacement for mica in electrical applications, hence the name 'for mica' (and nothing to do with ants). It is used more now for worktops.

[¶] The author spent many a happy hour as a food analyst at the bench testing for these and other food colours using simple paper chromatography.

5 1B, 2C, 3H, 4G, 5A, 6E, 7D, 8F
Additional notes:

- Charles Darwin: Beagle, or rather, HMS Beagle, whose second voyage carried the young naturalist around the world. He subsequently published *The Voyage of the Beagle*, his diary journals which brought him fame
- Albert Einstein: Bibo the parrot which was given to Einstein as a 75th birthday present. Since then an African Grey Parrot called Einstein has found some fame as The Talking Texan Parrot (hatched 1997 and available on YouTube)
- Edwin Hubble. Nicolaus Copernicus the cat. One astronomer honouring another! Hubble is perhaps better known for the Hubble Space Telescope.
- Nikola Tesla: Macak the cat. Tesla's childhood pet apparently got him interested in electricity, after stroking Macak produced 'a sheet of light' from static electricity, and of course, the rest is history (or perhaps science?)
- Thomas Huxley. Bulldog, or rather the sobriquet 'Darwin's bulldog' for his public support of Darwin
- Sir Ian Wilmut: Dolly the Sheep (named after the Country and Western singer Dolly Parton). Wilmut was the leader of the research group that cloned Dolly in 1996. Dolly (the sheep) was the first mammal cloned from an adult somatic cell.
- Sir Isaac Newton: Diamond the dog. According to some accounts, Diamond upset a candle and set fire to much of his owner's works.
- Ivan Pavlov: Drooling dogs and Pavlov's work on conditioned responses and digestion for which Pavlov was awarded the 1904 Nobel Prize in Physiology or Medicine.

6 B Spanish fly

7 Cinnabar Moths.
A. Thins. B Sarin. C Barns. D Mobs. E Scar

8 Brimstone. There is some conjecture that the yellow brimstone butterfly gave the name to 'butter-fly'. (There is also a brimstone moth).

9 Neon, Ne. The fish is the *Neon tetra*, so named because of its bright colouring. It is a native of streams in the Amazon basin.

10 A Graphite, a soft black form of carbon used in 'pencil leads' (which contain no lead at all). The name graphite is derived from the Greek *grapho* to draw.‖

‖ Derwent Pencil Museum, the home of the first pencil is a short walk from Keswick town centre in Cumbria.

11 A Shark. The word squalene derives from the Latin for shark, *squalus*. It used to be extracted from sharks' livers.

12 B 200 litres.**.

13 B Edward Jenner (1749–1823)

14 Aspirin

15 B Theobromine††.

Despite its name, there is no bromine in theobromine. The name derives from the scientific name of the cacao tree *Theobroma cacao*, the *Theobroma* being derived from the Greek *theo*, god and *broma*, food, so 'food of the gods.'

16 Phosphorus, P. The gaseous compounds are phosphine (IUPAC name phosphane) PH_3 and diphosphane (P_2H_4) which are spontaneously combustible in combination. Many other organisms exhibit phosphorescence (and bioluminescence) although the word phosphorescence does not always imply the causative agent is phosphorus. (The Greek word *phosphoros* means light-bringing and is the root of the element phosphorus and the word phosphorescence). Interestingly, the English marine biologist Peter Herring and author of the book The *Biology of the Deep Ocean* is known for his work on the camouflage, colouration and bioluminescence of animals of the deep. As for red herring, it is said that this saying was about using a red herring (*i.e.* a smoked herring which is a kipper) to put hounds off the scent when chasing a rabbit.

** Dr Thomas Beddoes (1760–1808) was a physician and chemist (and mentor of Humphry Davy). At his Pneumatic Institute in Bristol (which Davy took over in 1798), in addition to demonstrating the physiological effects of other gases (such as nitrous oxide), Beddoes believed that gases emitted by cattle might be a cure for tuberculosis (as butchers seemed to have a lower incidence of TB) and he piped bovine belches and flatulence into his patients' rooms (nice). Not so strange when you think about Edward Jenner's subsequent development of the world's first vaccine (for smallpox), derived from cows and their cowpox. (Jenner's wife Catherine unfortunately died from TB in 1815). The words vaccine and vaccination derive from Variloae vaccine ('smallpox of the cow').

†† The other three compounds listed are present in chocolate.

17 C Sauvignon Blanc. So next time someone says their wine tastes like cat's pee, they could well be telling the truth. Cat ketone also gives the characteristic odour of blackcurrant leaves and the descriptor 'blackcurrant leaf' is often used for Sauvignon Blanc.

18 C Vanadium, V. Japan once had a plan to harvest these creatures from their coastal shelf just for their vanadium. There are over a thousand species of sea cucumbers and they are sold as a delicacy in Asia and are also used in Chinese medicine.

19 A Brazilian pit viper. Captopril was approved for use in the US in 1981 and in Europe in 1984. Merck subsequently went on to develop Enalapril which became Merck's first billion-dollar-selling drug in 1988.

20 Chilli pepper

21 Pig iron is the crude iron obtained from the smelting process and took the form of ingots with a branching structure formed within sand, similar to a litter of piglets suckling on a sow.

22 Stinkbugs. (Apparently the flavour of the Mexican variety is a blend of cinnamon and mint).

23 Both B & C: tiny hairs on a gecko's feet help maximise contact with surfaces to allow the Van der Waals molecular forces to work.

24 Common Vampire Bat

25 Silk moth

Botany Bay

1 Juniper

2 You will find her first name in the word magnesium

3 1B, 2C, 3D, 4E, 5A

4 B Salicylic acid, after *Salix*, the Latin for willow.[††] A proprietary make of the painkiller, aspirin (acetylsalicyclic acid), was named as follows: *a* for acetyl and *spir* for Spiraea, the meadowsweet plant genus which by then was the source of the salicylic acid.

5 Foxglove. The active constituents are known as *digitoxin* and *digoxin*, the latter approved as an essential medicine by the World Health Organisation.

6 Iodine, I

7 A Ace. B Lace. C Lance. D Conceal. E Cochineal. F Nicholas Culpeper

8 B Explosives. The starch was used to produce acetone by fermentation for subsequent production of the explosive cordite. Previously the starch source was maize but German U-Boats could interrupt that supply. The cooker house (where the mash for fermentation was prepared) is on the site of the old Royal Naval Cordite factory in Dorset, England and is a listed monument (for more information and photos see www.historicengland.co.uk).

9 Geraniums

10 Kidney stones

11 A Aluminium

12 Stinging nettles sting through their hairs which are tipped with silica. The silica tips break off, and the natural needle releases a veritable cocktail of chemicals including formic, oxalic and tartaric acids and acetylcholine, histamine and serotonin.

[††] Note also that most cricket bats are made of willow. It still hurts if you get hit by one though!

13 The crocus, *Crocus sativa.* Saffron Walden is a market town in England, so named in the 16th century when the surrounding area's saffron production was at its peak.

14 The (opium) poppy, *Papaver somniferum*, the latter name having the same roots as insomnia, somniferous and somnambulation, from the Latin *somnus*, sleep. Morphine was discovered by a 21-year old apprentice apothecary Friedrich Sertürner (1783–1840) who named it after *Morpheus*, Ovid's name for the god of sleep.

15 Lavender. (For another way of preventing moth damage, see the opening page of the *Occupations* section).

Brain-Teasers

1 Silver 'transmutes' to quicksilver, *i.e.* mercury

2 Sodium chloride, **NaCl** in Bar**NaCl**es and potassium chloride, **KCl** in Sac**KCl**oth

3 Links! Li N K S (the elemental symbols)

4 Did you fall for that one? Nobelium, No is the most negative element! The most *electronegative* stable element is fluorine, F.

5 Heavy water D to O or D_2O (D standing for deuterium)

6 Their elemental symbols do not relate directly to the elements' names, rather to names in other languages:

- Antimony, Sb from Latin *stibnium*
- Iron, Fe from Latin *ferrum*
- Lead, Pb from Latin *plumbum*
- Gold, Au from Latin *aurum*
- Mercury, Hg from Latin *hydragyrum* meaning 'water-silver'
- Potassium, K from Latin *kalium*
- Silver, Ag from Latin *argentum*
- Sodium, Na from Latin *natrium*
- Tin, Sn from Latin *stannum*
- Tungsten, W from German *wolfram*

7 They all begin with letters of the Greek alphabet: beta, chi, pi, psi, rho, tau, respectively

8 The next number in the sequence is 85. The numbers are the atomic numbers of the Group 17 elements (the halogens), reading down from fluorine to astatine.

9 The numbers in the cells refer to the atomic numbers of the elements' symbols:

B	U	C	K	S
I		O		O
S	O	N	I	C
O		I		K
N	I	C	K	S

10 Tennessine (Ts), iodine (I), nitrogen (N) and the element reading down is tin (Sn).

11 Zirconium (Zr), iron (Fe), neon (Ne), cerium (Ce) and the element reading down is zinc (Zn).

12 Their elemental symbols may be formed from the first and last letters of each element.

13 The second element's first 2 letters are formed from the elemental symbol of the first element:

- Iron (Fe) Fermium
- Arsenic (As) Astatine
- Copper (Cu) Curium
- Gallium (Ga) Gadolinium
- Lanthanum (La) Lawrencium
- Nickel (Ni) Nihonium
- Radium (Ra) Radon

14 The second of each pair's elemental symbol is formed from the first and last letter of the first element:

- Carbon — Copernicium (Cn)
- Chlorine — Cerium (Ce)
- Francium — Fermium (Fm)
- Gold — Gadolinium (Gd)
- Fluorine — Iron (Fe)
- Iron — Indium (In)
- Silicon — Tin (Sn)

15 Norway is osmium. The elemental symbols of each of the elements may be formed from the first two letters of the respective capital cities of each country (**Br**ussels, **Ca**iro, **Li**sbon, **Na**irobi, **Os**lo).

16 The numbers refer to the atomic numbers of the elements. Put the corresponding elements together and you have A psi (ψ), B iota (ι), C beta (β), D chi (χ), E rho (ρ).

17

Z	I	N	C		Q			P	R	E	T	E	X	T
	N		O		U		N		H		R		E	
P	S	Y	C	H	E		U	T	O	P	I	A	N	S
	I		C		S		C		D		V		O	
	S	C	I	N	T	I	L	L	A	T	I	O	N	S
	T						E				A			
Y	E	L	L	O	W	C	A	K	E		L	I	F	E
	N						R						O	
E	T	C	H		V	I	D	E	O	G	A	M	E	S
			Y				E						T	
A	M	E	D	E	O	A	V	O	G	A	D	R	O	
	A		R		P		I		O		A		L	
J	U	R	A	S	S	I	C		W	I	L	L	O	W
	V		N		I		E		N		E		G	
B	E	A	T	I	N	G			S		K	E	Y	S

18 The sum of the atomic numbers of the elements in each compound is the same, 28.

19

I	R	I	D	I	U	M
R	H	O	D	I	U	M
C	O	D	E	I	N	E
A	N	I	L	I	N	E
P	E	N	T	A	N	E
R	H	E	N	I	U	M

20 The elemental symbols are Ca, Se, In and the compound is casein.

21 The elemental symbols are Cu, Ba, Ne and the compound is cubane.

22 The elemental symbols are Ni, Ac, In and the compound is niacin.

23 Lead, Pb. The words reading down are neon, moon, moon, moan, loan, load and lead.

24 Copper, Cu. The elemental symbols of the others are US State abbreviations (AL, Alabama; AR, Arkansas; CA, California; CO, Colorado; Florida, FL, Georgia, GA; Indiana, IN; Louisiana, LA; Maryland, MD; Minnesota, MN)

25 The elements (and their symbols) begin with the Roman numerals in sequence: I, V, X, L, C, D, M

Calling All Units

1 B (femtogram 10^{-15} g), E (nanogram 10^{-9} g), D (microgram 10^{-6} g), A (centigram 10^{-2} g), C (kilogram 10^{3} g).

2 1G, 2D, 3A, 4C, 5B, 6F, 7E

3 They are all anagrams of each other. Ohms, Mhos and Mohs respectively. Mhos are the reciprocal of Ohms, and the Mohs Scale of Mineral Hardness is named after the German mineralogist Friedrich Mohs (1773–1883).

4 A Caesium, Cs. The International System of Units defines the second as the duration of 9, 192, 631, 770 cycles of the microwave frequency of the spectral line corresponding to the transition between two hyperfine energy levels of the ground state of caesium-133.

5 B Radium, Ra, actually radium 226. The Curie was originally defined as 'the quantity or mass of radium emanation in equilibrium with one gram of radium (element).'

6 Townie. Notice. Cation. Atomic. Diatom. Domain. Monday.

7 A Hydrogen (H, Henry). B Fluorine (F, Farad). C Nitrogen (N, Newton). D Vanadium (V, Volt)

8 Astronomical Unit which is roughly the distance from the Earth to the Sun, equal to 150 million kilometres or 8.3 light minutes.

9 Platinum, Pt

10 Length (metre), time (second), amount of substance (mole), electric current (ampere), temperature (kelvin), luminous intensity (candela).

Compound Interest

1 Ask wh[ETHER] [I'M IN E]ngland. S[URE A]m! But a neighbour's in a bad s[POT AS H]e's being p[ESTER]ed. ETHER, IMINE, UREA, BUTANE, POTASH, ESTER

2 C Calcium oxide, CaO, lime.

3 C Silver nitrate, $AgNO_3$

4 Water, H_2O.

5 A CALCIUM CARBONATE, $CaCO_3$. B NITROUS OXIDE, N_2O. C SILVER NITRATE, $AgNO_3$. D CARBON TETRACHLORIDE, CCl_4. E COPPER SULFATE, $CuSO_4$. F SODIUM CHLORIDE, NaCl

6

C	5	H	5	N
5		F		H
H	G	1	9	4
1		7		C
2	8	4	F	L

7

C	2	H	4		M	C	5		B	A	T	1
5		3		C		1		1		L		8
H		P	B	3	O	4		S	U	M	4	1
1		O		P		H		2		I		W
1	0	4		O	S	1	9	2		T		
N						0		S		Y		N
O	M	E	G	A	3		W	1	A	1	A	A
2		M		L		C						2
		B		I	R	1	9	1		1	2	C
9		L		E		8		0		3		R
8	T	E	E	N		H	G	C	L	2		2
T		M		3		1		C		B		O
C	L	3	5		H	2	S		M	A	Y	7

8 1H, 2I, 3D, 4G, 5K, 6B, 7E, 8F, 9J, 10A, 11C

Additional notes:

Aqua regia Mixture of nitric acid and hydrochloric acid.

Note this is called aqua regia since it dissolves the king's metal, gold. Indeed, when Germany invaded Denmark in 1940, Hungarian radiochemist Georg Charles de Hevesy dissolved the gold Nobel Prizes of Max von Laue and James Franck in aqua regia to prevent the Nazis from taking them, and placed them on a shelf in his laboratory at the Niels Bohr institute. After the war, he precipitated the gold out of the acid, and the Nobel Society recast Franck and von Laue's awards from the original gold. Georg de Hevesy subsequently won his own Nobel Prize in Chemistry in 1943 'for his work on the use of isotopes as tracers in the study of chemical processes.'

Lunar caustic Silver nitrate

In early times, silver being inferior to 'the perfect' gold (which was associated with the sun) was associated with the moon. Alchemists represented it by the sign of a crescent moon and called it Luna (Latin). Silver salts were therefore called lunar salts.

Microcosmic salt Ammonium sodium phosphate

The microcosm refers to the small world of human nature as opposed to the macrocosm of the wider universe. The salt was so called as it was originally obtained from human urine.

9 A Galactose, Lactose. B Quinoline, Quinine. C Putrescine, Purine. D Butadiene, Butane

10 Indium, In

11 Epsom Salts, $MgSO_4{\cdot}7H_2O$. Epsom is of course famous for its annual flat horse race, the Epsom Derby.

12 *Arsole*, or more strictly speaking, according to IUPAC (International Union of Pure and Applied Chemistry, the ultimate arbiter of molecular monikers), 1*H*-arsole.

13 1,2-dimethyl chickenwire. It doesn't exist! Chemists are such wags aren't they!

14 1c, 2a, 3b, 4f, 5e, 6d

15 A Ran. B Rain. C Train. D Latrine. E Interval. F Irrelevant. G Silver nitrate

16 Silver carbonate. A Eros. B Site. C Orbs. D Steer. E Solver

17 Rust

18 Polychlorinated biphenyl

19 The compound reading down is anthracene. The others are: A Ammonia, B Niobite, C Toluene, D Heavy Water, E Raffinose, F Arginine, G Cysteine, H Ethane, I Nickeline, J Ephedrine. The quotation by the English radiochemist and Nobel Prize Winner Frederick Soddy (1877–1956) was:

As buildings are built of bricks so compounds can nowadays be built up out of atoms.

20 1 Roc, 2 Mop, 3 Pup, 4 Pie, 5 Red, 6 Car. The elements are copper and radium.

21 The PAH is phenanthrene, $C_{14}H_{10}$

22 The compound reading across is silicone.

A. Arsine, Basics, Hosier
B. Calico, Police, Relics
C. Bicorn, Recoin, Second
D. Linear, Renege, Tenets

23 The ore reading across is limonite.

A. Allies, Feline, Holier
B. Almond, Osmose, Remote
C. Canine, Conics, Genial
D. Astern, Cetene, Intend

24 Orthodox and paradox (obviously)

25 Fluorine, F. The compound, $C_{13}H_{10}$ is named fluorene, so-called as it shows a strong (violet) fluorescence. It is the basis of some dyes.

26 Fertiliser – a source of nitrogen. This is not the first time this compound has led to tragedy. In 1921, an explosion in a German ammonium nitrate factory in Oppau killed 561 people and 173 people were killed in the Chinese port of Tianjin in 2015 by an ammonium nitrate explosion.

27 The scientists reading across and down are: (Ludwig) Mond, (Robert) Bunsen, (Rosalind or Benjamin) Franklin, (James) Watson, (Hideyo) Noguchi, (William) Perkin, (Friedrich) Mohs

28 The forms or sources of energy reading across and down are coal, kinetic, windpower, paraffin, fission, potential and heat

29 The things you might find from a mine reading across and down are tin, horn silver, uraninite, arsenic, cinnabar, bornite and ore

Discoveries

1 B Radon, Rn

2 The chemist reading down is (Sir) Humphry Davy. Across clues: A Hodgkin, B Urey, C Magnesium (Mg), D Potassium (K), E Hafnium (Hf), F Rayleigh, G Young, H Dewar, I Americium (Am) (the other element was Curium (Cm)), J Viagra, K Yttrium (Y). The quotation is: '*We have discovered the secret of life*', was announced by Francis Crick on 28 February 1953 at the Eagle Inn in Cambridge about his and James Watson's elucidation of the double helix structure of DNA.

3 A Tantalum, Ta. The word 'tantalise' comes from the mythological story that led to the naming of the element.

4 C Sir William Ramsay (1852–1916). At the time, he was the only person to have discovered an entire periodic group of elements.

5 The light bulb

6 B Moscovium, Mc

7 Teflon

8 Warfarin. In 1951, the suicide was prevented by the administration of Vitamin K and prompted research into the poison's therapeutic use. Warfarin was developed by American scientist Karl Paul Link in 1948 but may be traced back to a cattle disease decades before where cows were excessively bleeding after dehorning. The reason? Fungus-infected sweet clover hay and the active compound identified in 1938 as dicoumarol. Link 'improved' on dicoumarol's deadly activity and named the new compound after the Wisconsin Alumni Research Foundation.

9 Gallium, Ga. Paul Emile Lecoq de Boisbaudrun named it from the Latin for Gaul (France). But 'le coq' is French for the rooster, and the Latin for rooster is *gallus*. (He apparently refuted this scurrilous claim).

10 B Celluloid. Schönbein's discovery made him rich and famous, and he was invited to demonstrate the new explosive at the Woolwich Arsenal in London where he presented Queen Victoria and Prince Albert with a brace of pheasants, the first ever shot with guncotton cartridges.

11 1C, 2B, 3F, 4E, 5D, 6A

12 Graphene. There is now a National Graphene Institute based at The University of Manchester, England, for which see https://www.graphene.manchester.ac.uk/ngi/

13 Photography. Daguerre invented the *Daguerrotype* after he'd left a copper plate coated with silver iodide in a cupboard after it had been exposed in a camera obscura. (Before this there was no image visible on the plate). After exposure to mercury fumes, it appeared. The French painter Paul Delaroche (1797–1856) declared '*From today painting is dead*' and cartoons showed engravers hanging themselves. Little did he know that modern painters such as David Hockney would one day themselves be painting on tablets or that digital artworks themselves would sell for millions of dollars using NFTs (non-fungible tokens).

14 Palladium, Pd

Earth Sciences

1 Late Cretaceous. A Slate. B Cores. C Ceres (Cerium). D Ores. E State

2 A ALLUVIAL. B ANISIAN. C HYDROGRAPHY. D MAGMA. E MINMI. F NEOGENE. G TENNANTITE

3 Iridium, Ir

4

P	Mo	No	Cl	I	N	I	C	H	I	Na	C	La	Y	
Cu	Ac		As	P	Ba	S	Al	Ti	C	La	V	As		
Te	B	H	S	Li			Ne	Ta				Ar		Ca
	N	I	Y	O		O		Ni		Es	Am	K	Rb	B
		Na	C	Ce	Ge			Te		K	Zn	O	Co	Ra
			N	Ne	P					Er	Ni	Se	Pr	C
	Ge			Ti	C	H	Al	K	Y	F		Zr	O	H
	O				Te		Al		Er				Li	I
Ar	P	Zn	U	Ca	Li	Te		O				S	Te	O
B	H						U		S		Ru		S	Po
Na	Y	Am	Mo	Ni	Te	S				Au				Ds
N	Si		B						S	Zn	Ru			
I	Cs	Ti	Ra	Er				I				S		
C	Re	Ta	Ce	O	U	S	Lv	Cl	I	No	C	La	Se	
						Si	C	Al	Ca	Re	O	U	S	

5 B Kernowite (after Kernow)

6 Ammonia, NH_3 (and also the ammonium ion NH_4^+). The fossil is ammonite. The god is Ammon. Horns of Ammon were curling ram's horns.

7 1A, 2F, 3D, 4C, 5B, 6E

8 The minerals are talc, chalk, calcite and alabaster:

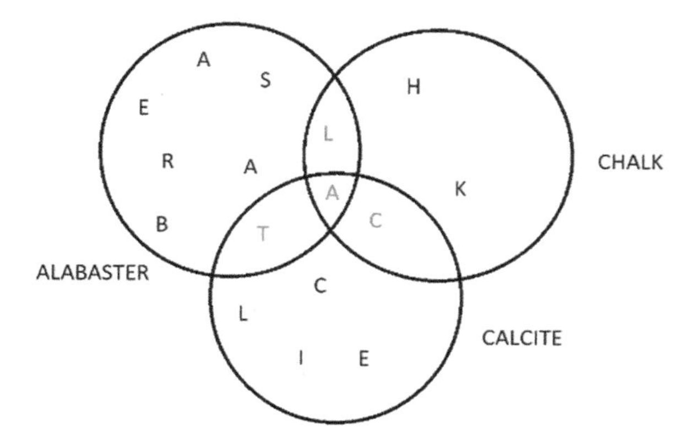

9 Jargon

10 B but you are allowed to believe C as well. This is the author's wife Angie on the Giant's Boot (the giant Finn McCool lost on his way back from the fight):

The columns (or *collonades*) were created by cracks propagating upwards during cooling of the lava and most are hexagonal due to the joints being 120° to each other, but some can have three to seven sides. The Giant's Causeway is now managed by the National Trust (see www.nationaltrust.org.uk/giants-causeway). When asked by his biographer James Boswell (1740–1795) '*Is not the Giant's Causeway worth seeing?*', English writer and lexicographer Samuel Johnson (1709–1784) is quoted as remarking '*Worth seeing, yes; but not worth going to see.*' The author disagrees.

11 The geological feature reading across is erratic. An erratic is a rock that is out of keeping with its geological surroundings, deposited there by a retreating glacier. The Giant's Boot is an erratic, deposited there at the end of the last Ice Age (about 10 000 years ago).

A. Alerts, Coerce, Sierra
B. Berate, Carafe, Curare
C. Attics, Notice, Patina
D. Lucre, Niche, Sects

12 Calcium carbonate, $CaCO_3$

13 Iron, Fe

14 Barium, Ba

15 C 4.5 billion years

Food & Water

1 C Quinine. If you shine UV light on a bottle of tonic water, it will fluoresce due to the quinine content.

2 Quark. (For more on this see the Particles chapter).

3 Calcium, Ca. Calcium is present in chalk as calcium carbonate ($CaCO_3$) and milk and cheese are good sources of the element.

4 Irn-Bru. Its original name was 'Iron Brew' until British legislation necessitated a tweak - the iron in Irn-Bru is a tiny amount (0.002%) of ammonium ferric citrate ($(NH_4)5[Fe(C_6H_4O_7)_2]$) which provides a faint taint of rust. The rusty colour of one of Scotland's favourite drinks (another one of course being whisky) comes not from rust but the food colours Sunset Yellow FCF (E110) and Ponceau 4R (E124). These colours are banned in some countries, and you'll find this statement on cans of Irn-Bru, in line with the UK FDA's guidance: 'May have an adverse effect on activity and attention in children'.[†§]

[†§] As well as a fiery taste, some of its adverts are pretty fiery too, fiery enough to be banned.

5 Ethylene (or ethene).

6 Hydrogen sulfide, H_2S. If you're lucky enough to have silver spoons, when eating your boiled egg (preferably not rotten!), they will tarnish with time due to the sulfur in the egg, which forms the black silver sulfide Ag_2S.

7 Hard and soft (water).

8 The condiment is salt, the permitted food sweeteners aspartame, sorbitol and sucralose:

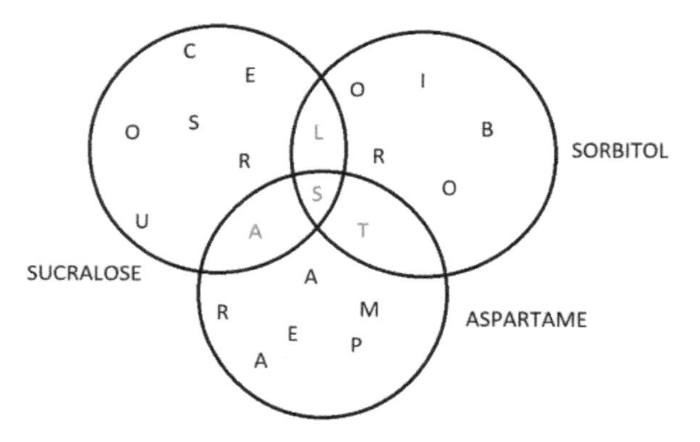

9 $CH_3CH_2OH(l) + O_2(g) \rightarrow CH_3COOH(aq) + H_2O(l)$
Ethanol Oxygen Acetic Acid Water

This reaction is what will turn wine into vinegar through oxidation (the word *vinegar* literally translates as sour wine from the Old French *vyn egre*). The word *acetic* derives from the Latin acetum, vinegar (from *acere*, to be sour).

10 B Benzoic acid (a food preservative). Did you think it was nicotinic acid? Surely so close to the word nicotine that it can't be a vitamin? Wrong! But we're being a bit crafty. The name nicotinic acid was used since this compound was first isolated from the tobacco plant and is otherwise now known as niacin (changed to this because of the addictive nature of nicotine and associated negative connotations) or vitamin B_3. The generic name of the tobacco plant *Nicotiana* (and the basis of the word nicotine) comes from Jean Nicot (1530–1604) who introduced tobacco into France in 1560. This predates 1586 when Sir Walter Raleigh supposedly introduced it to Britain (although it was probably used before this date).

11 C Sulfur, S which is somewhat ironic perhaps, as the latest (as of 2022) onion to reach the UK that is not such a tearjerker, is the Sunion.

12 Bird's Custard Powder.

13 B Cider. Not only was lead found in some of the cider presses' plumbing, but lead shot was used to scrub out the equipment between pressings. Once the lead was removed from the process, the colic disappeared.[†¶]

14 Hydrogen cyanide, HCN, produced from amygdalin (named from the ancient Greek for almond)

15 To produce activated charcoal for gas masks which protected against poison gas. (You can now buy flatulence filtering underwear employing activated charcoal).

16 A CAROTENE. B ACESULFAME K. C CARBON DIOXIDE. D BENZOIC ACID. E INDIGO CARMINE. F CELLULOSE. G CARRAGEENAN. H L-CYSTEINE. I ASPARTAME

17 OXO

18 Asparagus (and the amino acid is asparagine). The cause of the smell in urine after digestion is asparagusic acid which releases sulfur-containing compounds on digestion (including dimethyl sulfide, dimethyl sulfoxide, dimethyl sulfone and methanethiol). Thiol compounds are particularly smelly and the human nose can detect them at concentrations of a few parts per billion. Thiols are the compounds responsible for the smell of skunk spray.

The French writer Marcel Proust (1871–1922) wrote:

[†¶] The author was a Food Analyst in the 1980s for Somerset County Council, another county famed for its cider. He remembers with some fondness the many local ciders sent to the laboratory from across the county for chemical analysis. Apart from the routine alcohol content, arsenic analysis was also undertaken (as arsenicals were previously used agriculturally) with a nifty little 'wet chemical' technique. Zinc and hydrochloric acid were added to 50 ml of the cider in a conical flask, with a trap above it to catch any potential arsine gas (AsH_3) in pyridine imbued with a colorimetric reagent (silver diethyl dithiocarbamate) which would turn various shades of red dependent on the amount of arsenic (the production of arsine this way forms the first step in a similar (forensic toxicology) test for the element, known as the Marsh Test). He can't recall ever finding any arsenic, though a similar rosy facial colouring may also be produced by the ingestion of copious quantities of the beverage.

Asparagus '...transforms my chamber-pot into a flask of perfume.'

Another amino acid, aspartic acid was first discovered in 1827 by hydrolysis of asparagine and is also named after the vegetable. The sweetener aspartame is the methyl ester of the aspartic acid/phenyalanine dipeptide, was discovered in 1965 and again ultimately derives its name from the vegetable (with the 'me' ending for methyl ester).

19 Potato. The word solanine comes from the genus *Solanum* (for instance, potato is *Solanum tuberosum*). The green colouring is due to chlorophyll, not solanine, but is an indication of elevated levels of the alkaloid.

20 7-Up (which does not contain lithium citrate).

21 C Joseph Priestley (1733–1804). Priestley provided a method for making carbonated water for the crew of James Cook's second voyage to the South Seas, thinking it would prove a cure for scurvy. He did not exploit the commercial potential of fizzy drinks but a German-Swiss watchmaker and amateur scientist Johan Jacob Schweppe (1740–1821) did. The rest is history and Schweppes regards Priestley as 'the father of our industry.'

22 B Raffinose

23 Coca-Cola

24 Pufferfish. They are called pufferfish due to their defence mechanism of greatly inflating themselves with water which turns them into a spiny ball three or four times their normal size. Japanese chefs have to earn a licence after three years of training to be able to safely prepare the dish.[†‖]

25 A Capsaicin. Pepper spray was originally prepared as a defence against bears and other dangerous animals and is also known as bear spray.

[†‖] The explorer Captain James Cook had a lucky escape in 1774, surviving fugu poisoning, but a pig which ate the fish entrails did not.

26 B Dill and spearmint

Carvone is also found in caraway seeds (hence its name from the Latin *Carui carvi*). It is *d*-carvone that is in both caraway and dill. (Menthol, the active constituent of mint also exists in two optically active forms, but only one of them is found naturally).

27 A, B and C, although the majority of it comes from duck feathers and there are vegan alternatives produced by fermentation of plant-based raw materials

28 The heat of chillies. Scoville Heat Units (or SHUs) range from 8000 for jalapeno chillies to 16 000 000 for pure capsaicin.

29 Red cabbage. Solutions of the boiled cabbage turn red (acidic) to blue, green and yellow as the pH increases due to the presence of anthocyanin.[†**]. Popular in Thailand, the butterfly pea flower also changes colour with pH: blue at pH 7, purple with a drop of citric acid and magenta with another squirt, enabling colour-changing cocktails.

30 B Grapefruit

31 C Nutmeg

32 Quinoa. The Incas called it the 'mother grain' although quinoa is not a true grain (as it is not the seed of a grass plant).

33 C Rhododendron. Mad honey is used in small doses in Turkey as a folk medicine. Its toxicity is due to blocking of sodium channels in the body, and although poisoning in humans is rare, there are many reported cases of fatal poisoning in cattle or pets.

Greek warrior-writer Xenophon wrote of mad honey in 401 BC '*... but the swarms of bees in the neighbourhood were numerous, and the soldiers who ate of the honey all went off their heads, and suffered from vomiting and diarrhoea, and not one of them could stand up, but those who had eaten a little were like people exceedingly drunk, while those who had eaten a great deal seemed like crazy, or even, in some cases, dying men. So they lay there in great numbers as though the army had suffered a defeat, and great despondency prevailed. On the next day, however, no one had died, and at approximately the same hour as they had eaten the honey they began to come to their senses; and on the third or fourth day they got up, as if from a drugging.*'

[†**] The spice turmeric is also a pH indicator – yellow in acid, red in alkali.

34 Proof. 100° proof by the gunpowder test is equivalent to 57% ABV (alcohol by volume). The author has tested many alcoholic drinks for their alcohol content when he was a food analyst, distilling the alcohol off and then evaluating the ethanol content by specific gravity, or by a technique called gas chromatography. On the opposite side of the world in New Zealand, you can buy Smoke & Oakum Manufactory's Gunpowder Rum which actually has the three main components of gunpowder in it, namely saltpetre (potassium nitrate), charcoal and sulfur, although at 51.6% ABV it would not pass the gunpowder test!

In Transit

1 1G, 2A, 3H, 4I, 5C, 6D, 7B, 8F, 9E

2 C Vanadium, V

3 C Sulfur, S

4 In the car battery.

5 In the batteries of electric vehicles.

6 The airbag. Once the impact velocity is above a critical threshold, a sensor will activate an electrical impulse to detonate the sodium azide, thus releasing the nitrogen as a gas, inflating the airbag within 25 thousandths of a second:

$2NaN_3 \rightarrow 2Na + 3N_2$

7 Magnesium, Mg. Oscar Pereiro won the 2006 Tour de France on a magnesium bike.

8 In the catalytic converter fitted to the exhaust system. The precious metal is platinum, Pt although combinations of other metals may also be used.

9 Trike

10 The first Messerschmitt Bubble car (made 1953–1955). Previously Messerschmitt had been an aircraft manufacturer. The small car bears a passing resemblance to a cockpit. The KR in KR175 comes from the German *Kabineroller* ('scooter with cabin').

11 B Hydrogen, H_2. Both balloon flights took place in Paris, France.

12 Traditional school buses. They were painted with Chrome Yellow paint to make them visibly bright in gloomy conditions. However, it has since been replaced since actual Chrome Yellow ($PbCrO_4$) contains toxic lead.

13 Benzene, C_6H_6. For more on this molecule, see the Compound Interest chapter.

14 Mercedes Benzene

15 C 1958. Interestingly the German U-Boat U-234 was captured by the Allies enroute to Japan after the German surrender and was found to contain ten containers marked 'Japanese Army.' They contained 560 kg of uranium oxide. In addition, the U-boat was carrying armour-piercing shells and two Me-262 jet fighters.

16 Magnesium, Mg. This led to them also being called mag wheels. Nowadays aluminium is more likely to be used.

17 Elon Musk.[‡†]

[‡†] Note the word 'Musk' comes from the Sanskrit for 'testicle' since traditionally musk for perfumery was extracted from animal scent glands.

Litmus Tests

1 The pH scale used to measure acidity and alkalinity. It is logarithmic and inversely indicates the concentration of hydrogen ions (H^+) in solution:

$$pH = \log(1/[H^+])$$ where $[H^+]$ is the concentration of hydrogen ions in moles per litre.

It is thought that the p of pH stood for potential (and obviously the H for hydrogen) but it could be other things too such as the French *puissance*, German *potenz* or Danish *potens*, all meaning power or potential. (Sørensen published in all three languages). The Carlsberg Laboratory was also the birthplace of the *Kjeldahl* technique for the analysis of nitrogen and protein in foods, named after the Danish chemist Johan Kjeldhal (1849–1900) who developed it in 1883, as a way of determining the protein content of the malt used in the brewery.

At 25 °C, solutions with a pH less than 7 are acidic, and solutions with a pH greater than 7 are basic. (Sorensen was nominated for a Nobel prize in chemistry and physiology and medicine for his creation of the standardized scale, but never won).[‡‡]

2 A Sodium (Na), B Potassium (K), C Rubidium (Rb), D Strontium (Sr)

3 B Accurate and imprecise

4 B High performance liquid chromatography, although Hewlett-Packard does make HPLC (and other chromatographic) equipment. (Other suppliers are available).

5 A, B, C & D are all correct

6 Arsenic, As

7 A Thin layer chromatography

8 A Nicotine. The criminal case became quite famous and in the early 1990s. The Brussels Comédie Claude Volter showed the play *Nicotine et guillotine*, based on this case.

---

‡‡ Beer tends to be slightly acidic, about 4 on the pH scale.

9 A Syringe. B Petri Dish. C Autoclave. D Microscope. E Centrifuge. F Liebig Condenser

10 A Die. B Diet. C Tired. D Credit. E Accredits. F Desiccator

11 Conical Flask. A Coals. B Fins. C Skin. D Flanks. E Cons. F Cocks. G Slack. H Sick

12 The answers are Buchner flask, litmus paper, Bunsen burner, spatula, precision balance:

B		U	C		H		N	Er
	Fl			As		K		
		Li		Tm	U		S	
	Pa				P			Er
	B		U	N		Se	N	
		B		U		Rn	Er	
	S		P		At		U	La
P		Re	C	I	S	I	O	N
Ba		La		N		Ce		

13 A Despite Bouguer having originally clocked the effect, the law was named the Beer–Lambert Law after August Beer (sounds like a summer ale but he was actually a German chemist, physicist and mathematician) and the Swiss-French polymath Jean-Henri Lambert. The law describes the attenuation of light to the properties of the material through which it is travelling.

Occupations

1 A Naphthalene $C_{10}H_8$:

Percival Potts (not to be confused with the madcap inventor Caractacus Potts of *Chitty Chitty Bang Bang* fame) published his observations on the high rate of scrotal cancers in London chimney sweeps and postulated a link between soot and the observed cancers. These sweeps were usually children who started their life of grime at eight years of age, although some 'apprenticeships' could start as young as four. Potts' observations are believed to be the first to establish a reliable cause and effect relationship between an environmental agent and cancer. One hundred years later, von Volkman showed a link with skin cancer among workers in the German coal tar industry. Direct experimental evidence for Potts' results were forthcoming in 1915.

2 Mercury, Hg. Mercury nitrate was used in the felting of hats, and the resulting condition was known as *erethism* (also known as *hatter's shakes* and *mercury madness*). Beaver and rabbit furs were dipped in a solution of mercury nitrate and then dried to get the short hairs of the fur to mat together. This process was known as *secretage*, apparently since it was known originally only to a few French workmen in the 17th century, or *carroting* since it turned the felts orange.

3 Lead, Pb after the Latin *plumbum.* And, of course, plumbing, which at least in previous years used lead pipes and lead-based solder to join them.

4 Matches. The Matchgirl's Strike of 1888 involved the women and girls who worked at the Bryant & May factory in London in Bow, London. In 1891 the Salvation Army opened another match factory using the less dangerous red phosphorus and paying better wages. The story of the strike was made into a musical called *The Matchgirls* which premiered at the Globe Theatre in 1966, one of the numbers being 'Phosphorus.'

5 Painting luminous dials.

6 C Fingerprinting. The 'grey powder' used in dusting the fingerprints was a mercury concoction, known as *hydragyrum cum creta*, or 'mercury with chalk', prepared by rubbing the liquid elemental mercury with the chalk until the globules were no longer visible.

7 Beryllium, Be. Chronic beryllium disease affects the lungs and is also known as berylliosis.

8 Asbestos. Diseases associated with exposure to asbestos include asbestosis, mesothelioma and lung cancer. The American actor and racing driver Steve McQueen (1930–1980) suffered from mesothelioma which he thought was a result of his exposure while removing asbestos insulation from pipes on a ship while he served in the Marines (although he could have been exposed to it from other sources, such as the protective suits and helmets he wore as a racing driver). McQueen actually died of a heart attack following surgery to remove tumours.

9 Helium, He

Particles

1 The electron.

2 B No. It's not cheese (for which see the Food & Water chapter). The correct answer is charm.[‡§]

3 The Higgs boson. The LHC and the search for the Higgs boson also feature in Robert J Sawyer's novel *Flashforward.*

4 A Hyperon. B Antilepton. C W-Boson. D Nucleon

5 The subatomic particles are bottom quark, gluon, meson and muon:

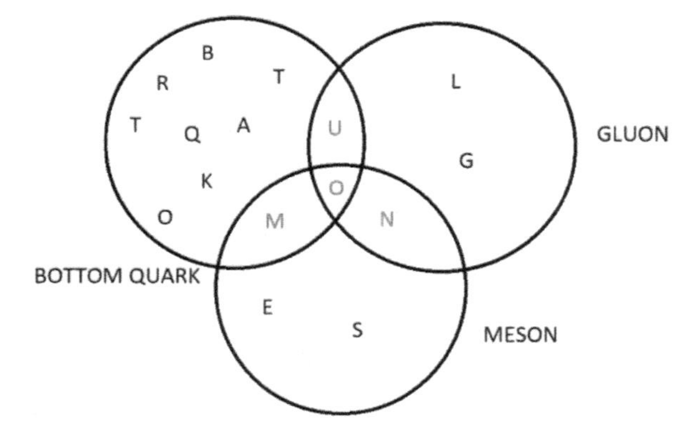

[‡§] For anyone interested, *Quark, Strangeness and Charm* was the seventh studio album by the English rock group Hawkwind, released in 1977.

6 The subatomic particles are positron, fermion, lepton and pion:

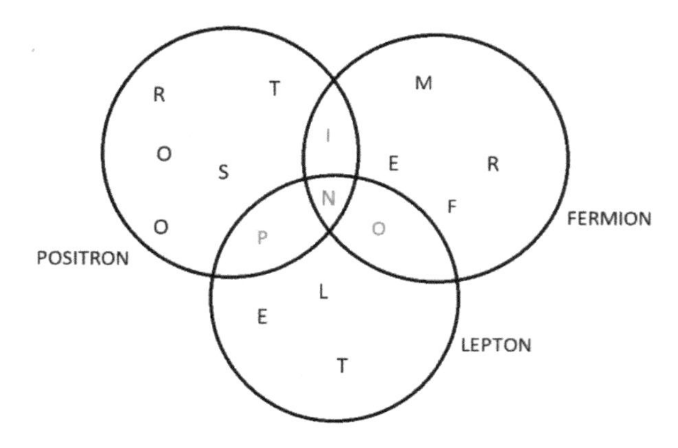

7 Fermium, Fm, named after Enrico Fermi (1901–1954) who also had the fermion named after him.

8 Electrons

9 The subatomic particle reading down is strange quark. The other answers are A Spin, B Tachyon, C Rubidium, D Alpha, E Neutrino, F Gamow, G Electron, H Quantum, I Ultraviolet, J Atom, K Rutherford, L Kaon. The quotation is '*If I could remember the names of all these particles I would have been a botanist.*' Enrico Fermi.

10 C Plum pudding. This particular model of the atom was disproved with Ernest Rutherford's experiments firing alpha particles at a thin metal foil-most of the particles passed straight through (as most of the atom is empty space), but some struck the nucleus and bounced straight back. Rutherford is quoted as saying it was '*as if you fired a 15-inch shell at a piece of tissue paper and it came back and hit you.*'

11 The World Wide Web.

12 C *Finnegan's Wake* by James Joyce:

– Three quarks for Muster Mark!
Sure he hasn't got much of a bark
And sure any he has it's all beside the mark.

13 Cyclotron, invented by US physicist Ernest Orlando Lawrence (1901–1958) for which he won the 1939 Nobel Prize in Physics.

14 A The Great Pyramid of Giza in Egypt

15 C Pion (or pi-meson)

16 *Quod erat demonstrandum*, Latin for that which was to be demonstrated, often found at the end of formal proofs.

17 B Neutrino. Fermi was Italian and Fermi's name for this particle meant 'little neutral one' and a neutrino is a type of fermion (a subatomic particle named after Fermi, with spin of ½). It has since been discovered that there are three different types of neutrinos: electron neutrinos, tau neutrinos and muon neutrinos. And each neutrino also has its antiparticle, an antineutrino.

18 A Positron Emission Tomography

19 A ALICE, an acronym for one of the LHC's experiments: *A Large Ion Collider Experiment* which is dedicated to heavy-ion physics. It is designed to study the physics of strongly interacting matter at extreme energy densities in a hot, dense particle soup (known as a *quark-gluon plasma*) which mimics what was happening a few millionths of a second after the Big Bang. Alice of course was Lewis Carroll's most famous character, appearing in his book Alice in Wonderland.

20 Beer. Believe it or not, Cellarmaker Brewing Company makes a Bubble Chamber beer.

Image courtesy of Cellarmaker Brewing Company

Periodicity

1 Tungsten, W

2 Molybdenum, Mo

3 Bromine, Br and osmium, Os from the Greek *bromos*, stench and the Greek *osme*, smell.

4 C Oxygen, O_2, originally called *dephlogisticated air* by the English chemist Henry Cavendish after the word *phlogiston.* The so-called *Phlogiston Theory* posited that all combustible bodies contained a fire-like element, phlogiston which was lost on burning. This theory was later debunked, particularly by the French chemist Antoine Lavoisier.

5 Seaborgium, Sg and praseodymium, Pr

6 Scandium, Sc

7 B Tin (Sn) although some historians dispute the Napoleonic story. The crumbling of the tin solder that once held Captain Scott's food cans together on his fated Antarctic venture is another tragic story potentially linked to this element. The effect is also known as 'tin leprosy'.

8 Elements are:

- Chlorine, Cl; krypton, Kr; rhodium, Rh and the middle one iron, Fe:

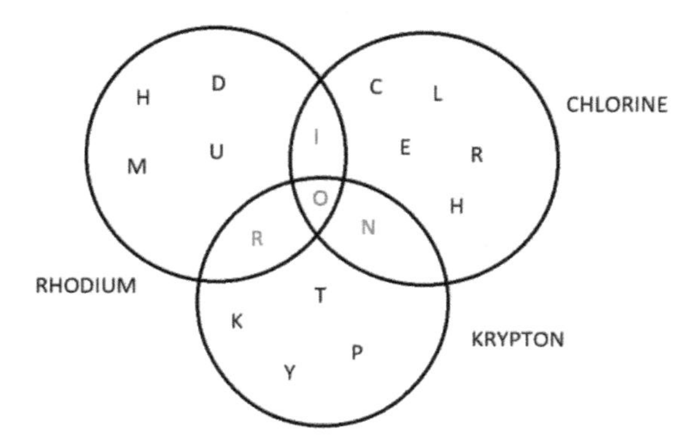

- Gadolinium, Gd; lawrencium, Lr; molybdenum, Mo and the middle one lead, Pb:

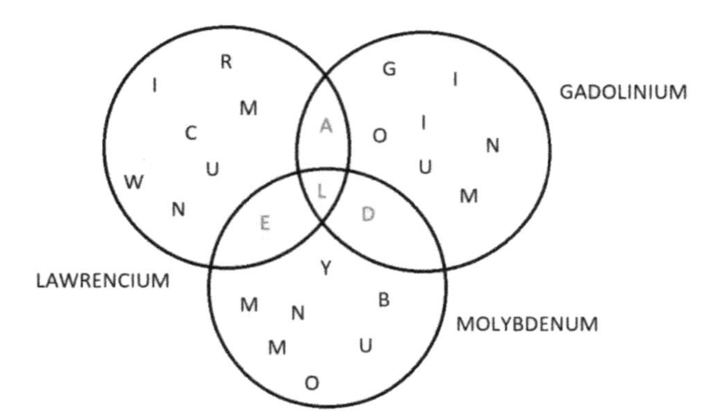

9 The completed periodic table is below and the five elements are arsenic (As), tantalum (Ta), titanium (Ti) and neon (Ne) with astatine (At) being the fifth element formed by the elemental symbols of the previous four.

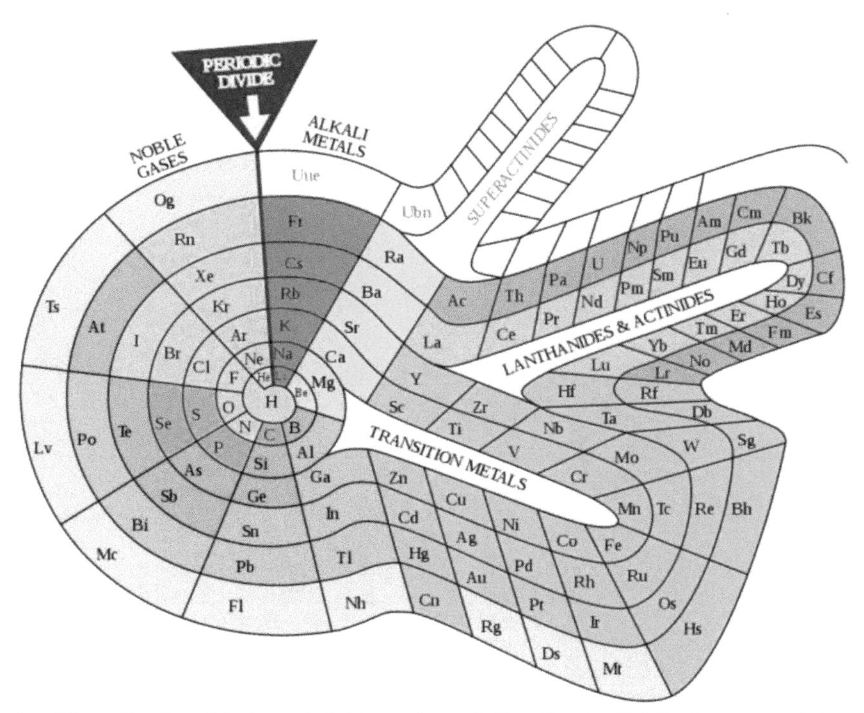

10 Platinum, Pt

11 Aluminium (or aluminum if you're American), Al

12 Gallium, Ga. A gallium teaspoon will seemingly disappear when used to stir the hot beverage (for which see *The Disappearing Spoon...* in the Bibliography)

13 Caesium (cesium in American English), Cs

14 Cobalt, Co. The great scientist Michael Faraday (1791–1867) claimed Marcet's *Conversations on Chemistry* had played a vital role in his development.

15 Silver, Ag

16 Platinum, Pt

17 Osmium, Os and tungsten, W (known in German as wolfram whence the element gets its symbol W), so OSRAM combines the first two letters of osmium and the last three letters of wolfram

18

Ne	W	T	O	N	S	T	H	Ir	D		La	W	O	F
	He		V		U		At		Es		N		K	
P	At	T	Er	N	S		C	H	I	P	O	La	Ta	S
	S		Po		Ta		He		C		Li			
Ca	T	Ni	P		I	N	T	Er	C	O	N	Ne	C	T
	O		U		Ne		S		At				Ar	
	Ne	W	La	N	D	S		T	Es	T	T	U	B	Es
	Br		Te								Es		O	
F	I	S	S	I	O	N		Ra	B	Bi	T	O	N	
	D				F		B		U		I		I	
Ar	Ge	N	T	I	F	Er	O	U	S		Mo	Ti	F	S
			O		P		Ta		H		N		Er	
Pu	B	Li	C	At	I	O	N		I	S	I	D	O	Re
	Al		Si		S		I		D		Al		U	
Mo	Ti	O	N		Te	Na	C	I	O	U	S	Ne	S	S

19 A Protactinium, Pa. B Berkelium, Bk. C Flerovium, Fl. D Ytterbium, Yb. E Lawrencium, Lr. F Californium, Cf. G Manganese, Mn. H Praseodymium, Pr. I Gadolinium Gd

20 A Cerium, Ce. B Iron, Fe, Francium Fr. C Sodium, Na. D Gold Au, Indium In. E Radium, Ra, Osmium, Os, Sodium, Na. F Tin Sn, Actinium, Ac, Titanium, Ti. G Tin Sn, Iron, Fe, Neon, Ne, Nitrogen N. H Rhodium, Rh, Thorium, Th. I Barium, Ba, Erbium, Er. J Neon, Ne. K Yttrium, Y, Terbium, Tb, Erbium, Er. L Iron, Fe, Zinc, Zn

21

B	U	N	Se	N
I		I		O
S	Ta	C	K	S
Ho		He		He
P	I	S	Te	S

22 Mendelevium, Md after Dmitri Mendeleev (1834–1907)

23 A Silver (Ag), B Iron (Fe), C Gold (Au), D Lead (Pb)

24 A Eon. B Neon. C Tenon. D Neutron. E Neutrino. F Roentgenium

25 1C, 2D, 3B, 4A

26 A Chlorine, Cl. B Nickel, Ni. C Hassium, Hs. D Actinium, Ac. E Francium, Fr. F Neodymium, Nd. G Terbium, Tb

27 Carbon

A. Diamond
B. Graphite (used in pencils and lubricants, its name coming from the Ancient Greek to write or draw)
C. Graphene-a single carbon layer thick
D. Buckminsterfullerene C_{60} (or 'Buckyball'), named after the US architect Richard Buckminster Fuller (1895–1983), widely known for his geodesic domes, hence the connection

28 Magnesium, Mg.

29 A BASIC. B CARBON. C GARBAGE. D NEUTRINO. E FLUORINE. F LATTICE. G POLECAT. H PROTONS

30 A HOLMES. B SCORSESE. C SCARY. D SERRATE. E HEROINE. F NECTAR. G SELENITE. H GAWAIN

31 1B, 2C, 3D, 4E, 5F, 6G, 7A

32 The elements are reading down: niobium (Nb), and reading up thulium (Tm):

- N atriu M
- I in-sit U
- O kap I
- B rune L
- I mprompt U
- U ta H
- M andelbro T

33 The elements are reading down: lutetium (Lu), and reading up samarium (Sm), and livermorium (Lv) in the first clue:

- L ivermoriu M
- U mam I
- T a U
- E ule R
- T esl A
- I odofor M
- U re A
- M ean S

34 Europium, Eu

35 1729. This number is also known as the Hardy–Ramanujan Number after a visit by another mathematician G H Hardy to Ramanujan then in his hospital bed in 1919, the year before his untimely death:

"I remember once going to see him when he was ill at Putney. I had ridden in taxi cab number 1729 and remarked that the number seemed to me rather a dull one, and that I hoped it was not an unfavorable omen. "No", he replied, "it is a very interesting number; it is the smallest number expressible as the sum of two cubes in two different ways."

Such numbers have since been branded 'taxicab' numbers by mathematicians, with the symbol Ta(n) or Taxicab(n), being defined as the smallest integer that may be expressed as the sum of two positive integer cubes in n ways, so 1729 is Ta(2) or Taxicab(2)[‡¶].

36 1 Acidic, 2 Raceme, 3 Badger, 4 Cobalt, 5 Casino, 6 Nitric. The central element is carbon.

37

S	O	C	K	S
C		O		O
U	N	W	Ra	P
F		Er		P
F	U	S	S	Y

‡¶ Ramanujan's life and relationship with Hardy is brilliantly portrayed in the 2015 film *The Man Who Knew Infinity*.

38 The solvent is water. The scientists are (Rachel) Carson, (Laura) Bassi, (Harold) Urey. The kitchen implements are knives.

H	W	At	Er	C	Ar	S	O	N	B	As	Si	U
Re	Y	K	Ni	V	Es	F	I	Ts	Ta	P	Te	Ds

C	O	W	P	At	S		Re	S	P	I	Re	S
At		I		Ta		S		K		Re		I
I		F		I	N	H	Er	I	Ta	N	C	Es
O	N	I	O	N		I				I		Ta
Ni			At			F	O	S	Si	C	K	
C	O	P	Y	I	S	Ts		As			I	
	At			O				S			W	
	H			Ni		P	S	Y	C	H	I	C
	S	Ta	C	C	At	O			O			O
Ar		H				U		B	O	S	O	N
Re	V	I	S	I	O	N	Ar	Y		O		C
Te		N		O		Ds		Te		Ni		H
S	K	I	N	N	Y		P	S	Y	C	H	Es

39 1C, 2A, 3E, 4B, 5D

40 Tin (Sn) and lead (Pb)

41 Lead (Pb)

42 C Technetium (Tc)

43 C Ruthenium, Ru (albeit alloyed with a little iridium). The letters RU may be found on the nibs and 'reimagined' Parker 51s were introduced in 2021.

44 Holmium, Ho

45 Dysprosium, Dy whose name comes from the Greek for 'hard to get', in allusion to the difficulty in the element's detection and isolation.

46

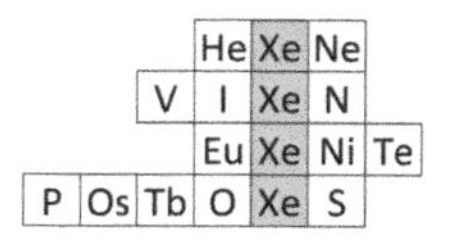

47

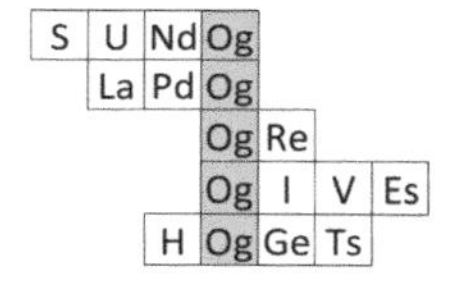

Pigments of the Imagination

1 The rainbow and its colours from the first letter of each mnemonic: red, orange, yellow, green, blue, indigo, violet

2 Xanthophyll (after the Greek for yellow *xanthos*). There is some speculation that the artist Van Gogh's 'yellow period' could be a result of medication he was prescribed (*digitalis*) which gave a yellow tinge to his vision (known as *xanthopsia*).

3 They are all named after colours.

4 Urine

5 The mordant is alum, the compounds mauveine, melanin and ultramarine:

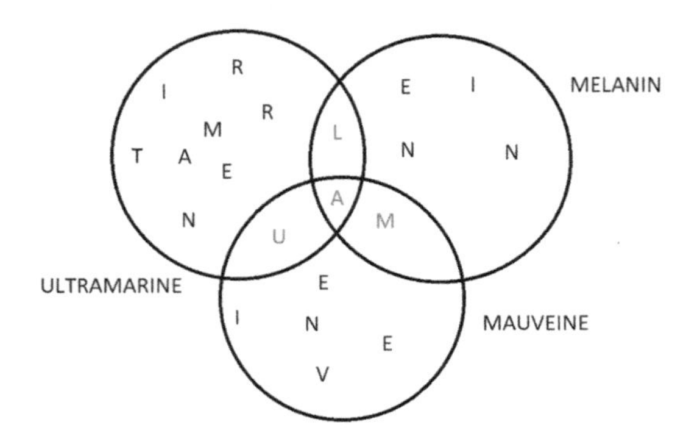

6 Here are two possible solutions:

1. WOAD	WOAD
2. GOAD	GOAD
3. GOAT	GLAD
4. GOUT	GLAM
5. GLUT	GLUM
6. GLUE	GLUE
7. BLUE	BLUE

7 B Chromium (Cr), the pigment being chrome yellow.

8 B Iron oxide. Red, yellow and brown pigments come from iron oxides.

9 Flamingo. The common green algae, *Haematoccus algae* is one type of algae which contains a high astaxanthin content. Note farmed salmon and captive flamingos have synthetic astaxanthin added to their feed to compensate for the lower astaxanthin content of their diet.

10 1E, 2B, 3D, 4A, 5C

11 1C, 2A, 3E, 4D, 5B

12 Cannabis

13 C Carbon nanotubes. The name Vantablack comes from 'vertically aligned nanotube arrays' and the colour black.[‡‖]

Places

1 A2, B6, C5, D4, E3, F1

The odd one out is silver, as Argentina is named after silver (Latin *Argentum* whence the element gets its symbol Ag), while all the other elements are named after the places.[‡**]

2 They are all flags of places that have elements named after them: (left to right on top row, then left to right on bottom row) californium (Cf) – California, europium (Eu) – Europe, francium (Fr) or gallium (Ga) – France, germanium (Ge) – Germany, nihonium (Nh) – Japan, polonium (Po) – Poland, ruthenium (Ru) – Russia, tennessine (Ts) – Tennessee.

3 Silicon, Si

4 Magnesium Mg and manganese Mn, after *Magnesia*, a former prefecture of Greece. The place name also gives rise to the word magnet, magnetic and magnetism due to the prevalence of ore that attracts iron.

5 Nihonium, Nh, first synthesised in Japan in 2004. However, Japanese scientist Masataka Ogawa thought he had discovered a new element in 1908 (thinking it was element 43) and called it nipponium. Element 43 was later synthesised in a particle accelerator and named technetium, Tc.

6 Sulfur (the French for this element being *soufre*). Soufrière is located within the caldera of the dormant Qualibou volcano and the area is geothermally active. Tourists can visit the volcanic site, as has the author. He still remembers the sulfurous stench.

[‡‖] The mention of the writer Philip Pullman is a reference to his trilogy of books '*His Dark Materials*'.

[‡**] Additionally, the river that forms part of the border between Argentina and Uruguay, Rio de Plata, translates as River of Silver.

7 Mercury, Hg

8 Rhenium, Re. The Rhine, although the Spanish did originally call platinum *platina del Pinto* (little silver of the Rio Pinto, near where it was first found)

9 Boron, B. The salt/ghost town is Borax. The medical dressing is boracic lint (sometimes spelt brassic), the rhyming slang, skint.

10 Brussels. Iron, Fe.

11 Yttrium, Y. The village is Ytterby (which translates as 'outer village') and is found on the Swedish island of Resarö. The other elements named after Ytterby are erbium, Er, terbium, Tb and ytterbium, Yb. Four other elements can trace their discovery to the Ytterby minerals, namely holmium, Ho (after Stockholm), scandium, Sc (after Scandinavia), thulium, Tm (after the mythical name for Scandinavia), and gadolinium, Gd (after Johan Gadolin (1760–1852)). Several of Ytterby's streets are named after these elements (such as Gadolinitvägen, Terbiumvägen, Yttriumvägen, and Ytterbystandsvägen) and others (such as Tantalvägen – after tantalum) and also minerals (such as Glimmervägen, Glimmer being Swedish for mica, Stenvägen, Sten being Swedish for stone).

12 C Swansea. There is also a Copperopolis in California (so named because of its copper mine).

Romans mined for copper in North Wales, where ore deposits were shallow. For a while, the Cornish then had the edge on copper production, until a huge deposit was discovered on Parys Mountain on Anglesey in 1768. For many years copper was smelted in the Swansea area since they had easy access to coal there, but not without dire consequences on the surrounding environment. It was said of the area that 'nowhere in Britain is there a more dismaying example of man creating wealth while impoverishing his environment.' Although Swansea's surgeon and registrar argued that the smoke had prophylactic properties.[§†] In the heart of what was Swansea's Copperopolis is a Nature Reserve called White Rock and the old spoil heap is now a haven for rare plants and butterflies including Bee orchids and the Small Blue butterfly. In 2022 Copr Bay (Copper Bay) opened to the public in Swansea as a celebration of the cities copper connections. This includes a copper/gold coloured bridge dubbed the 'Crunchie' due to its resemblance to the honeycombed bar.

[§†] The great English chemist and physicist Michael Faraday (1791–1867) visited Parys Mountain in 1819 on his tour of Wales and noted that the copper smelters were less polluting than those of Swansea but that the iron-rich water from the precipitating pits and the rivers into which it ran looked like blood, staining the rocks in the local Amlwch harbour. Amlwch now has a Visitor Centre celebrating its industrial heritage, The Copper Kingdom. (See https://copperkingdom.co.uk/).

13 Promethium, Pm (Rome)

14 C China

15 Copper, Cu. In Roman times, copper was mined principally on Cyprus, the origin of the name of the metal, from *aes cyprium* (metal of Cyprus), later corrupted to *cuprum* (Latin).

16 Gallium, Ga from the Latin *Gallia* for France, discovered by French chemist Paul-Émile Lecoq de Boisbaudran (1838–1912) in 1875 and francium, Fr, discovered in 1939 by the French physicist Marguerite Perey (1909–1975). (Perey was a student of Marie Curie).

17 Aluminium, Al

18 Greece. The name *Verdigris* comes from the Old French for 'green of Greece.'

19 Plaster of Paris.

20 A Gin. B Sing. C Singe. D Singer. E Sterling. F Burlington House.

21 A Lavoisier Island is in Antarctica.

22 1E, 2G, 3C, 4A, 5B, 6D, 7F

23 B Iron, Fe

24 France and Switzerland.

25 Helium, He

26 Lead shot (the lead being alloyed with variable amounts of tin, antimony and arsenic to control the hardness). The lead was dropped from a great height into water to produce the shot. This method was replaced in the 1960s by the *Bleimeister method* which drips molten lead from small orifices about an inch into a hot liquid and then rolled along an incline, then dropped another three feet. The Cheese Lane Shot Tower now houses an office complex called *Vertigo.*

27 Titanium, Ti. The building is home to no fewer than 33 000 half-millimetre thick titanium panels.

28 Copper, Cu. The lightning conductor did contain copper though (albeit after the original iron one glowed red after a strike).

29 1E, 2C, 3B, 4F, 5A, 6D

In addition, Colorado has many ghost towns with chemically connected names associated with its mining heritage including Calcite, Carbonate, Copper City, Crystal, Galena, Gold Hill, Iron City, Silver Creek, Tincup, Tungsten, Vanadium and Uravan (Uravan being a contraction of Uranium and Vanadium).

30 Polonium, Po, discovered by wife and husband team Marie and Pierre Curie. Marie was born in 1867 in Warsaw (which was Russian at the time). Discovered in 1898, she named the element after her homeland before Poland finally got its independence after the First World War.

31 C Astatine, At. The other elements' symbols are the two-letter international country codes for each of the other countries.

Planets

1 The answers are A Brown Dwarf, B Lockyer, C Aquarius, D Caroline, E Krypton, F Hydrogen, G Oxygen, H Lavoisier, I Eddington. The additional word reading down is Black Hole. The quotation is '*How inappropriate to call this planet Earth when it is quite clearly Ocean.*' Arthur C Clarke (1917–2008), sci-fi and science writer.

2 Tellurium, Te, from the Latin for Earth (*terra*), and Selenium, Se, from the Greek for moon (*Selene*).

3 Plutonium, Pu (and the heavenly body was Pluto).

4 A Burbidge. Margaret Burbidge (1919–2020). B Burney. Venetia Burney (1918–2009). C Fleming. Williamina Fleming (1857–1911). D Lise Meitner (1878–1968)

5 A DMITRI MENDELEEV. B GLENN T SEABORG. C SVANTE ARRHENIUS. D LINUS PAULING. E SIR HUMPHRY DAVY. F JOHN DALTON. G ROBERT BOYLE.

6 Phosphorus, P

7 Mercury. Completed in 2003, it has since been decommissioned.

8 Palladium, Pd. The asteroid was Pallas (minor-planet designation: 2 Pallas).

9 Fraunhofer Lines. A Liners. B Linens. C Hones. D Ferns. E Fauns

10 Helium, He. Helium constitutes about 24% of the mass of the Sun, a product of the hydrogen nuclear fusion.

11 The element is cerium. The heavenly bodies are Pluto, Earth, Neptune, Ganymede, Mercury, Saturn.

12 The stargazer is the astronomer Nicolaus Copernicus, obtained by replacing each of the elements in the cryptogram by the element immediately before it in the Periodic Table. The element is Copernicium, Cn.

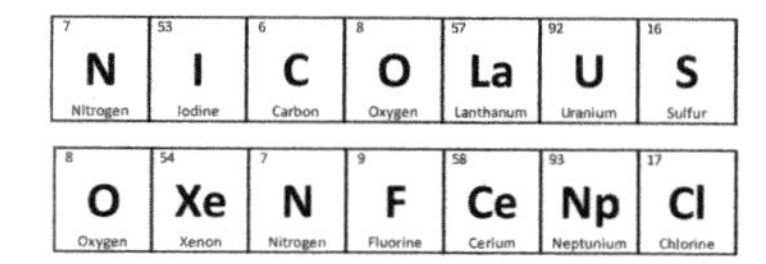

6	8	15	68	7	53	6	92	16
C	O	P	Er	N	I	C	U	S
Carbon	Oxygen	Phosphorus	Erbium	Nitrogen	Iodine	Carbon	Uranium	Sulfur
7	9	16	69	8	54	7	93	17
N	F	S	Tm	O	Xe	N	Np	Cl
Nitrogen	Fluorine	Sulfur	Thulium	Oxygen	Xenon	Nitrogen	Neptunium	Chlorine

13 B B2FH or B^2FH was a seminal paper on the stellar origin of the elements. *Synthesis of the Elements in Stars* was published in *Reviews of Modern Physics* in 1957. The moniker B^2FH comes from the first letters of the surnames of the authors, namely and Margaret Burbidge, Geoffrey Burbidge, William A Fowler and Fred Hoyle.

14 Hydrogen, H

15 Xenon, Xe. (XIPS stands for Xenon Ion Propulsion System).

16 C (Sir) Arthur Eddington (1882–1944)

17 Orbiting Saturn. Methone is a small moon that was discovered in 2004, although it wasn't photographed in detail until 2012 (by the Cassini spacecraft).

18 Lithium, Li

19 Germanium, Ge

20 Beryllium, Be

Poisons

1 A Arsenic, As[§‡]

In Bradford, over 200 people were poisoned and 21 victims died as a result of the adulterated sweets. The event prompted the passing of the Pharmacy Act 1868 and food legislation regulating adulteration. (The poisoning occurred due to then quite common practice of adulterating the sweets with what was cheap gypsum (also known as Plaster of Paris). However, a mistake at the pharmacy mistakenly provided arsenic trioxide instead of gypsum).

Further information on the beer case can be found at the FIRST REPORT OF THE ROYAL COMMISSION APPOINTED TO INQUIRE INTO ARSENICAL POISONING FROM THE CONSUMPTION OF BEER AND OTHER ARTICLES OF FOOD OR DRINK (6 July 1901).

2 Antifreeze. Much stricter wine laws were passed in Austria as a result of this scandal.

[§‡] *A is for Arsenic, The Poisons of Agatha Christie* is also a book, for which see the Bibliography.

3 Lead, Pb. High levels of lead, probably coming from the lead solder on the canned food they ate. The ratio of lead isotopes in the cans matched those in the exhumed bodies and differed markedly from the lead isotope ratios in the local Inuit population.

Beethoven's Hair covered the suspected lead poisoning of the composer.

4 1D, 2H, 3C, 4F, 5E, 6B, 7G, 8A

5 1F, 2C, 3A, 4D, 5B, 6E

6 A Gel. B Glen. C Galen. D Galena. E Analgesic. F Sparkling Cyanide

7 A Phosphorus, P (actually white phosphorus P_4)

8 C Strychnine

9 The answers across are beryllium, ethanol, lead, litharge, aconitine, digitalis, opium, nickel, Newton, atropine and reading down is belladonna. The quotation is '*Here's your arsenic dear and your weedkiller biscuit*' by Dylan Thomas from '*Under Milkwood*'

10 Potassium chloride, KCl. By a twist of fate, the first woman to be executed in the State of Arkansas since 1845 received a lethal injection of KCl on 2 April 2000 after she had killed her two children, one of them with an injection of KCl.

11 A Cyanide and mercury. B Lead and strychnine. C Polonium and ricin. D Cadmium and antimony.

12 Thallium, Tl

Radioactivity

1 Alpha α, beta β and gamma γ. These terms were coined by Ernest Rutherford (1871–1937).

Alpha particles (also called alpha rays) comprise of two protons and two neutrons bound together and are the equivalent of He^{2+} particles, *i.e.* helium nuclei. When an element undergoes alpha-decay, it shifts its position two places to the left on the Periodic Table.

Beta radiation comprises energetic electrons, with conversion of a neutron into a proton and the electron, so the element involved in this type of transmutation shifts one place to the right on the Periodic Table.

Gamma rays have the shortest wavelength and therefore the highest energy of any rays in the electromagnetic spectrum. Gamma decay normally occurs from a daughter nucleus formed from alpha or beta decay, emitting excess energy from its excited state.

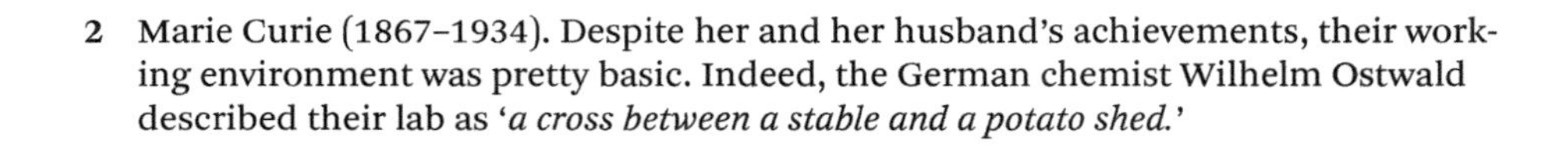

2 Marie Curie (1867–1934). Despite her and her husband's achievements, their working environment was pretty basic. Indeed, the German chemist Wilhelm Ostwald described their lab as '*a cross between a stable and a potato shed.*'

3 ^{12}C and ^{4}He

4 Fission

5 A Caesium-137. Despite her discovery, Margaret Melhase was blocked from pursuing a PhD. She did however work on the Manhattan Project.

6 Technetium, Tc

7 C Czech Republic. Hotel Radium Palace was built in the spa town of Jáchymov (then called Joachimstahl) and was the world's first radium spa. Marie Curie herself visited the town in 1925 and in 1898 with her husband Pierre who discovered radium in pitchblende from the town's mine.

8 Ernest Orlando Lawrence (1901–58). The Lawrence brothers collaborated in the field of nuclear medicine and treated their mother who was diagnosed with uterine cancer with X-rays. She had been given 3 months to live. She survived another 15 years (John later showed that neutron beams were more effective as a treatment). In 1936 John administered a dose of radioactive phosphorus to a 28-year-old leukaemia patient. This isotope was produced by the cyclotron and was the first time a patient had been treated with a radioisotope produced by a cyclotron.

9 An isotope of hydrogen, ^3H. A hydrogen atom that has two neutrons and one proton in the nucleus. Tritium is used in watch dials as a replacement for the more radioactive radium so the dials can be seen in the dark, the gas being housed in small tubes lined with a phosphorescent material, producing a luminous glow that can last decades through beta-decay of the tritium.

10 Radium, Ra. Twain also mentions polonium, Po in this story, also discovered by the Curies in the same year, and actinium, Ac which was discovered a year later in 1899 (supposedly by the French chemist André Louis Deberne (1874–1949) although this is disputed) from the pitchblende residues left by the Curies.

11 Cold fusion. The results could not repeated by other researchers and cold fusion has yet to make any dent in the world's energy budget.

12 Uranium, U

13 Zirconium, Zr

Scientists

1 1C, 2D, 3E, 4F, 5G, 6H, 7I, 8J, 9K, 10L, 11O, 12P, 13Q, 14T, 15S, 16B, 17A, 18M, 19N, 20R

2 1D, 2F, 3G, 4C, 5B, 6H, 7E, 8A

3 Roentgenium, Rg

4 Bohrium, Bh after Niels Bohr (1885–1962) and hafnium, Hf (*Hafnia* being the Latin name for Copenhagen). Hafnium was also discovered in Copenhagen in 1923 by Dirk Coster (1889–1950) and Georg von Hevesy (1885–1966).

5 Dubnium, Db. The previously proposed names for element 105 were nielsbohrium or hahnium.

6 Darmstadtium, Ds. The previously proposed names for element 110 were hahnium or becquerelium.

7

A. 2 Olivia Newton-John (b 1948)
B. Charles Darwin (1809–1882)
C. Aldous Huxley (1894–1963). Eric Arthur Blair (1903–1950) wrote as George Orwell and his dystopian novel was '1984.' The sci-fi writer H G Wells (1866–1946) was critical of Brave New World, describing it as '*a blasphemy against the religion of science.*'
D. The Curies. Marie Curie (1867–1934) received the prizes in Physics (in 1903) and Chemistry (in 1911). Her husband, Pierre Curie (1859–1906), shared the 1903 Physics prize with her. Their daughter, Irène Joliot-Curie (1897–1956), received the Chemistry Prize in 1935 together with her husband Frédéric Joliot-Curie (1900–1958). In addition, the husband of Marie Curie's second daughter, Henry Labouisse (1904–1987), was the director of UNICEF when he accepted the Nobel Peace Prize in 1965 on UNICEF's behalf.

8 Zinc, Zn

9

1. e Pythagoras. The square of the hypoteneuse is equal to the sum of the squares of the opposite sides (of a right-angled triangle).
2. a This is the Bragg Equation explaining X-ray diffraction with λ denoting the wavelength. The author's alma mater, Leeds University in 2021 opened the Sir William Henry Bragg Building, home of the Faculty of Engineering and Physical Sciences. Sir William Henry Bragg held the Cavendish chair of physics at Leeds from 1909–1915 where he worked on X-rays.

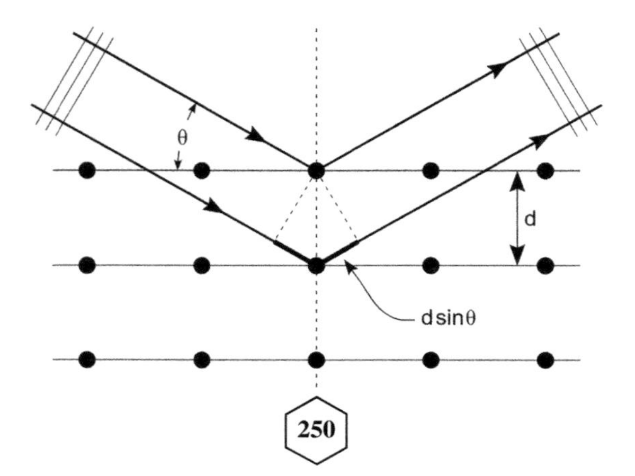

3. d Erwin Schrödinger. The Schrödinger equation describes the wave function of a quantum-mechanical system. For what it means, see Jim Al-Khalili's *Quantum, A Guide for the Perplexed.*
4. b Einstein of course. $E = mc^2$ is probably the most famous equation of all, explaining the interchangeability of Energy E and Mass m. The square of the speed of light c goes some way to explaining the power unleashed by the nucleus of the mighty atom.
5. c Newton's Law of Universal Gravitation

10 B They are all buried at Westminster Abbey. Surprisingly, Hawking was never awarded the Nobel Prize.

11 Meitnerium, Mt named after Lise Meitner (1878–1968). The epitaph was written by her nephew Otto Frisch (1904–1979) who was the co-discoverer of fission with his aunt. Frisch later worked on the Manhattan Project to produce nuclear weapons but Meitner refused to do so.

12 Curium, Cm. Pierre and Marie Curie.

13

A Dmitri Mendeleev (1834–1907) and Gregor Mendel (1822–84)
B Linus Pauling (1901–94) and Wolfgang Pauli (1900–58)
C Elizabeth Blackwell (1821–1910) and Joseph Black (1728–99)
D William Buckland (1784–1856) and Linda B Buck (born 1947)
E Srinivasa Ramanujan (1887–1920) and C V Raman (188–1970)
F George Brandt (1694–1768) and Hennig Brand (c 1630- c 1692 or c 1710)
G Bruno Rossi (1905–94) and Sir James Clark Ross (1800–63)
H Otto Fritz Meyerhof (1884–1951) and Viktor Meyer (1848–97)

14 The remaining unused cells spell out Newlands Octaves. The English Chemist John Newlands (1837–98) devised a system of elemental classification of the elements, seeing repetition in properties when ordered in increasing atomic weight 'like the eight note in an octave of music.'

	No.		No.		No.		No.		No.		No.		No.		No.
H	1	F	8	Cl	15	Co & Ni	22	Br	29	Pd	36	I	42	Pt & Ir	50
Li	2	Na	9	K	16	Cu	23	Rb	30	Ag	37	Cs	44	Os	51
G	3	Mg	10	Ca	17	Zn	24	Sr	31	Cd	38	Ba & V	45	Hg	52
Bo	4	Al	11	Cr	19	Y	25	Ce & La	33	U	40	Ta	46	Tl	53
C	5	Si	12	Ti	18	In	26	Zr	32	Sn	39	W	47	Pb	54
N	6	P	13	Mn	20	As	27	Di & Mo	34	Sb	41	Nb	48	Bi	55
O	7	S	14	Fe	21	Se	28	Ro & Ru	35	Te	43	Au	49	Th	56

Above: Newlands Table of the Elements. A few years later, Mendeleev published his Periodic Table.

S	**Ne**	F	F	Ga	Ar	H	C	N	I	W	Er	Ca
O	W	La	V	O	I	S	I	Er	In	Es	Th	K
Re	Ar	Er	Y	**W**	Th	Ga	N	K	At	Er	Ni	C
N	B	Re	N	O	In	Li	Er	Cr	I	**La**	Co	I
S	U	**N**	Pr	Er	K	P	Po	Ne	N	**Ds**	La	Cr
O	Rg	La	F	N	He	P	Al	B	Se	Er	U	S
Re	K	N	Ra	**O**	I	I	O	Ra	N	I	S	I
N	Be	F	**C**	H	Ce	Rn	Se	He	U	V	Co	C
Se	Nd	La	U	Ra	B	As	Si	N	B	Cu	P	N
N	At	Ta	I	Cl	Au	Si	U	S	Be	Es	Er	Ra
Ag	Ne	S	Ar	B	Er	Y	Re	O	Fl	Rg	Ni	F
Ta	I	Ar	Ne	T	I	Se	Li	U	S	O	Cu	**V**
N	I	K	O	La	Te	S	La	Mo	Nd	Ge	S	**Es**

15 Margaret Thatcher, née Roberts (1925–2013)

16 A American nuclear scientist Albert Ghiorso (1915–2010) is credited with co-discovering these elements. Ghiorso almost had an element named after him, *ghiorsium* (Gh), Element 118 which is now called oganesson after the Russian nuclear physicist Yuri Oganessian (b 1933).

17 They all won two Nobel Prizes, as follows:

- John Bardeen (Physics 1956, Physics 1972)
- Marie Curie (Physics 1903, Chemistry 1911)
- Linus Pauling (Chemistry 1954, Peace 1962)
- Frederick Sanger (Chemistry 1958, Chemistry 1980)
- K Barry Sharpless (Chemistry 2001, Chemistry 2022)

18 A Gertrude B Elion (1918–1999). B Emilie du Chatelet (1706–1749). C Hedy Lamarr (1914–2000). D Jennifer Doudna (b 1964)

19 C Technetium, Tc

20 1D, 2E, 3A, 4F, 5B, 6C, 7G, 8H

A few additional notes:

Antoine Lavoisier (1743–94)

Lavoisier revolutionised chemistry. Yet it was the French revolution that did it for him. In addition to his scientific pursuits, he was an aristocrat and tax collector, so perhaps no surprise that he lost his life to the guillotine.

Archimedes (c287–212 BC)

Known amongst many other things for his Eureka! Moment and the Archimedes Screw, he was killed when the Romans took Syracuse in 212 BC.

Giordano Bruno (1548–1600)

A supporter of the Copernican (heliocentric) system, Bruno was arrested by the Inquisition in 1592 but refused to recant after a protracted trial and was burned at the stake.

Marie Curie (1867–1934)

Awarded two Nobel Prizes and discoverer of the elements polonium and radium, Marie Curie died of leukaemia which could have been caused by her exposure to radiation during her work. Leukaemia also took the life of her daughter Irène Joliot-Curie, again possibly due to her radiation exposure during her scientific work (for which she too was awarded a Nobel Prize) and during her work as a radiographer during the First World War.

Nicolas Léonard Said Carnot (1796–1832)

Founder of thermodynamics, his seminal paper was *Reflections on the Motive Power of Fire* (1824) relating to the efficiency of steam engines. He is eponymized in the phrase *Carnot Cycle*. Carnot was a victim of the 1832 Paris cholera epidemic and many of his belongings and scientific writings were buried with him due to chlolera's contagious nature.

Pierre Curie (1859–1906)

Husband of and co-worker with Marie, he died in Paris. Crossing the busy Rue Dauphine in the rain, he slipped and fell under a heavy horse-drawn cart. One of the wheels ran over his head, fracturing his skull and killing him instantly.

Thomas Midgley (1889–1944)

A polio victim, Thomas Midgley devised a harness to help him out of bed. However, he became entangled and strangled himself. Midgley must go down as one of the biggest environmental miscreants ever. In 1921 he discovered that adding tetraethyl lead (TEL, chemical formula $Pb(C_2H_5)_4$) to petrol boosted its performance, and, in 1923 it was introduced for public consumption. Organic forms of lead are far more toxic than inorganic forms and Midgley himself spent several weeks fighting a severe illness related to his exposure. The following year, several workers who helped produce the additive in refineries in Ohio and New Jersey fell sick and died. Global contamination followed as a result of TEL's introduction, with lead levels in Arctic snows measuring 300 parts per trillion in the late 1970s. Lagging behind the USA, the EU finally banned the use of leaded petrol in 2000. In the very same year that he discovered TEL, Midgley attempted to find a safer refrigerant than the unpleasant sulfur dioxide and ammonia (which had a habit of occasionally leaking out and killing people: indeed a tragic accident that wiped out a Berlin family even prompted Albert Einstein and Leo Szilard to invent a fridge with fewer moving parts). Only three days after his TEL discovery, Midgley discovered that the CFC (chlorofluorocarbon) Freon-12 (CF_2Cl_2) made an excellent refrigerant. When he presented Freon to the American Chemical Society, he breathed in the gas and then exhaled it gently over a candle flame, which went out, demonstrating that his new compound was both non-toxic and non-flammable. Freon was widely used as a refrigerant and Einstein's fridges never made it to the shops. It was another 50 years before the effects of CFCs (which were also used in aerosols, foams *etc*) on our ozone layer were discovered.

William Whewell (1794–1866)

Whewell was quite the polymath (poet, historian of science, philosopher, priest and scientist-the word scientist itself being coined by Whewell). He died in Cambridge as a result of a fall from his horse. He was buried in the chapel of Trinity College, Cambridge.

21 Mendelevium, Md, named after the inventor of the Periodic Table, Dmitri Mendeleev (1834–1907).

22 Einsteinium, Es after Albert Einstein (1879–1955).

23 A Aid. B Idea. C Ideal. D Detail. E Elucidate. F Émilie du Châtelet

24 Weizmann became the first President of Israel in 1949.

25 Agnesi, silica, carbon, Onsager, Erwin, insulin, invar, Arne Tiselius, usnic, Ichneumon Wasp, sphalerite, tennessine, neoprene, neogene, nebula, Laura Bassi, silicone rubber, erucic, ichnite, Tesla, Laplace, cetene, Newlands Octaves, essonite. The astronomer from an anagram of the highlighted letters was Nicolaus Copernicus.

Ag	As	W	N	Mo	U	Ne	H	C	I
Ne	P	I	S	As	B	Ra	U	La	N
Si	H	Li	La	P	La	S	Te	U	S
Li	Al	C	Ce	Es	V	Ta	I	B	U
Ca	Er	O	Te	S	Te	C	N	Ne	Li
Rb	I	Ne	Ne	O	Ni	O	H	Ge	Se
O	Te	Ru	W	La	Nd	S	C	O	Ti
N	N	B	B	Er	U	C	I	Ne	Ne
S	Ne	S	S	I	Ne	O	P	Re	Ar
Ag	Er	W	I	N	S	U	Li	N	V

26 The scientist in the intersections is Bohr, after whom bohrium was named. The other scientists are Klaproth and Seaborg:

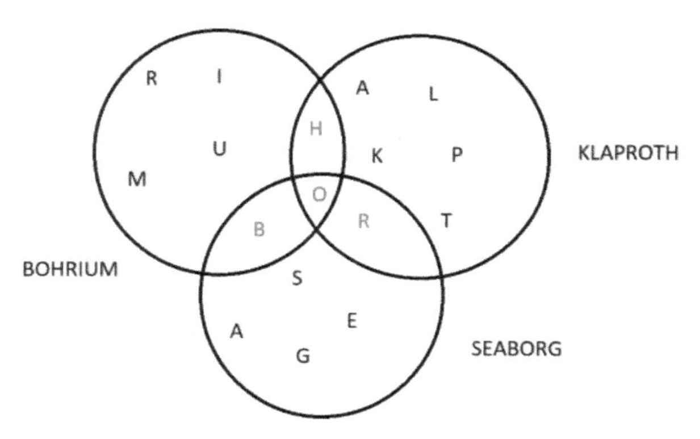

27 A Otto. B Wallace. C Martin. D Abel. E Arnold. F Franklin. G Solomon. H Benedict

28 1F, 2H, 3G, 4B, 5E, 6D, 7A, 8C

29 The chemist is Antoine Lavoisier (1743–1794). The elements are A Arsenic (As), B Nickel (Ni), C Tantalum (Ta), D Oxygen (O), E Iodine (I), F Neon (Ne), G Erbium (Er), H Lithium (Li), I Astatine (At), J Vanadium (V), K Osmium (Os), L Iridium (Ir), M Sulfur (S), N Iron (Fe), O Europium (Eu), P Radon (Rn). The quotation by the mathematician Joseph-Louis Lagrange (1736–1813) was:

Only a moment to cut off that head and a hundred years may not give us another like it.

30 1E, 2A, 3C, 4D, 5B

1. $E = h\nu$
   Max Planck. A photon's energy is equal to its frequency multiplied by the Planck's constant *h*.
2. $F = k_e(Qq/r^2)\hat{e}_r$
   Charles-Augustin de Coulomb, Charles-Augustin de Coulomb. Coulomb's Law is an electrostatic equation.
3. $R = m_e e^4/8\varepsilon_0{}^2 h^3 c$
   Johannes Rydberg. The Rydberg Constant, R relates to the electromagnetic spectra of an atom.
4. $S = k \log W$
   Ludwig Boltzmann. This equation forms the basis of the Second Law of Thermodynamics, with S representing entropy, a measure of a system's disorder. (This equation is also inscribed on Boltzmann's tombstone in Vienna).
5. $V = H_0 D$
   Edwin Powell Hubble. Hubble's Law describes the observation that galaxies are moving away from Earth at speeds proportional to their distance.

Superpowers

1 Krypton, Kr. The superhero Superman was born on the planet Krypton and his Achilles heel was the material Kryptonite. The name of Superman's dog was Krypto (what else?). Krypton is real. Kryptonite is fictional.

2 B Helium, He.

3 1B, 2E, 3D, 4C, 5A

4 Coca-Cola

5 A Helium, He

6 1D, 2E, 3C, 4B, 5F, 6G, 7A

7 C Superheating

8 The superheavy elements are:

A. Seaborgium, Sg (with an extra letter L)
B. Tennessine, Ts (with an extra letter O)
C. Rutherfordium, Rf (with an extra letter G)
D. Oganesson, Og (with an extra letter D)
(And of course, the other element is gold, Au)

9 A Pry. B Prey. C Ropey. D Poetry. E Entropy. F Kryptonite

10 Captain Universe. A Septa. B Sever. C Verse. D Peruse. E Casein. F Inverse

11 A Methane, CH_4

12 A Sparticles (imagine a cartoon of bosons and fermions crying out 'I'm Sparticle!'). *The Sparticle Mystery* is a children's sci-fi TV series for CBBC where an experiment at a Large Hadron Collider-like facility goes wrong. Its final episode was filmed at the ISIS Neutron and Muon Source at the Harwell Science and Innovation Campus in Oxfordshire, UK. (*Sparticle* is also the pseudonym of a puzzle compiler for the *New Scientist* magazine).

The Body

1 Iron, Fe

2 Iodine, I. Iodine deficiency in the diet is counteracted by the addition of it to common salt (as sodium iodide or potassium iodide). Lack of iodine can also lead to mental retardation, as pointed out by the British philosopher and Nobel Laureate Bertrand Russell (1872–1970): '*The energy used in thinking seems to have a chemical origin; for instance, a deficiency of iodine will turn a clever man into an idiot. Mental phenomena seem to be bound up with material structure.*'

3 Mercury, Hg

The mercury compound (calomel, Hg_2Cl_2) killed *Treponema pallidum*, the organism that caused syphilis, but sometimes killed the patient, with intense sweating and salivation being one of the tell–tell side effects (indeed, the poet John Donne (1572–1631) referred to a 'cold quicksilver sweat' in his poem *The Apparition*).

4 Folic acid, the name deriving from the Latin *folium* (leaf) in reference to the discovery of the compound in spinach leaves. Peanuts and broccoli are also a good source.

5 Lactic acid, $CH_3CH(OH)COOH$

6 Nitrous oxide, N_2O.

7 Arsenic. The 1940 biographical film *Dr Ehrlich's Magic Bullet* is about the Nobel Prize winning German physician and scientist, and pioneer of chemotherapy who found a cure (based on arsenic) for syphilis. The compound, arsphenamine was marketed under the tradename *Salvarsan* (from the Latin "salv(o)" to save, "ars" for arsenic plus and "an" an arbitrary suffix).

8 1H, 2B, 3A, 4G, 5F, 6C, 7E, 8D

9 A Glutamine, B Isoleucine, C Proline, D Asparagine, E Cysteine, F Methionine

10 The essential amino acids are methionine, threonine and tryptophan, and the essential element is iron, Fe:

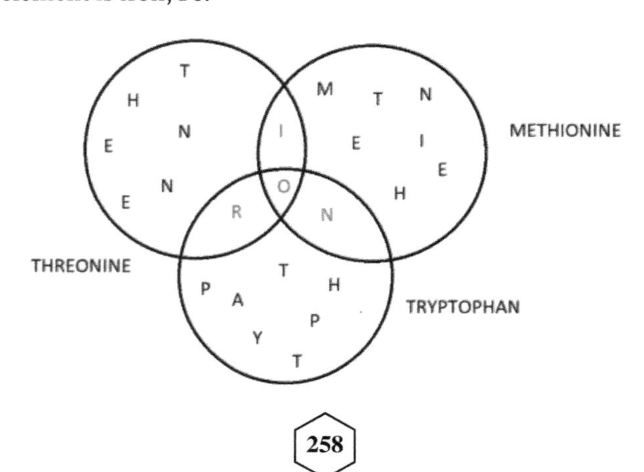

11 Uric acid, $C_5H_4N_4O_3$, although its IUPAC name is 7,9-dihydro-1*H*-purine-2,6,8 (3*H*)-trione. Phew! Uric acid is produced when the body breaks down substances called purines. Purine-rich foods include anchovies, beer, liver, and mackerel.

12 Chloroform. In 1848, the first death due to chloroform anaesthaesia was recorded (Hannah Greener). Edwin Bartlett, a grocer, also died as a result of chloroform poisoning on New Year's Day 1886. Analysis of the dead man's inflamed stomach showed that a large dose of chloroform had been swallowed. However, Adelaide Bartlett, his nineteen-year-old wife, was acquitted of murder on the grounds that she had bought the chloroform only to repulse her husband's unwanted advances.

13 Antimony, Sb

14 We now know 'carbolic acid' as 'phenol'. Lister pioneered its use as an antiseptic in 1865. Phenol was isolated from coal tar in 1834, and indeed coal tar formed the feedstock of the fledgling Victorian chemical industry (for which also see the Chapter Pigments of the Imagination). The mouthwash *Listerine* is of course named after the 'father of modern antisepsis' (as are the food-borne bacteria *Listeria monocytogenes*).

15 Arsenic, As

16 Cadmium, Cd. (The author can confirm the meaning of *Itai*, having heard it from the mouth of one of his grandchildren, Kiki, whose mother is Japanese). In the 1984 Japanese film *The Return of Godzilla,* cadmium shells are fired into Godzilla's mouth, knocking him unconscious.

17 John Dalton (1766–1844). Colour-blindness is also called *Daltonism.* Dalton thought he viewed his world through a blue filter so instructed his assistant Joseph Ransome to cut open an eyeball and look at the vitreous humour (the gelatinous substance inside the eyeball) but it was clear. The assistant also made a hole on the other eye and looked through it to see if colours appeared normal. They did. What remained of the eyeballs was preserved in a jar and left with the Manchester Literary and Philosophical Society until 1995 when a group of Cambridge physiologists examined the genes in the retinal cones for the three forms of colour-blindness. Dalton was a *deuterope* with a defect in the middle-wavelength optical pigment.[§§]

[§§] Due to his Quaker beliefs, Dalton refused a doctorate degree from Oxford University since he thought he would have to wear scarlet robes (mistaking the colour due to his colour-blindness). They were actually green robes so he could begrudgingly accept the honour.

18 Albert Einstein (1879–1955).

19 Calcium Ca, Carbon C, Hydrogen H, Nitrogen N, Oxygen O, Phosphorus P.

20 Chlorine Cl, Magnesium Mg, Potassium K, Sodium Na, Sulfur S

21 Silver, Ag (the clue's in the word argyria, from the Latin for silver *argentum*, which gives rise to its elemental symbol)

The Two Cultures

1 Rhodium, Rh

2 1G, 2B, 3D, 4A, 5H, 6C, 7F, 8E, 9J, 10I, 11K

3 A Platinum Blonde (1931). B Arsenic and Old Lace (1943). C Carbon Copy (1981). D Plutonium Baby (1987). E Mercury Rising (1998). F The Calcium Kid (2004). G My Zinc Bed (2008). H Iron Man 3 (2013). I The Neon Spectrum (2017)

4 Platinum, Pt

5 A Nickelback. B Lead Belly. C Silver Bullet Band. D Goldfrapp. E Freddie Mercury

6 1f, 2d, 3a, 4e, 5c, 6b

7 B Hydrofluoric acid, HF. The acid eventually ate through the bottom of the bath and the floorboards below.

8 Mercury, Hg

9 Borodin was a classical composer. Another Russian classical composer Nikolay Rimsky-Korsakov (1844–1908), wrote this about Borodin (1833–87):

Borodin was an exceedingly cordial and cultured man, pleasant and oddly witty to talk with. On visiting him I often found him working in the laboratory which adjoined his apartment. When he sat over his retorts filled with some colourless gas and distilled it by means of a tube from one vessel to another, I used to tell him that he was 'transfusing emptiness into vacancy'[§¶]*. Having finished his work, he would go with me to his apartment, where he began musical operations or conversations, in the midst of which he used to jump up, run back to the laboratory to see whether something had not burned out or boiled over; meanwhile he filled the corridor with incredible sequences from successions of ninths or sevenths.*

10 1E, 2C, 3G, 4H, 5F, 6D, 7A, 8B

11 Matt Groening, American cartoonist and creator of *The Simpsons* and *Futurama. Frinkonium* is named after the Springfield Professor Frink in *The Simpsons* (and Frink receives a Nobel Prize). *Jumbonium* is an element with atoms large enough to be seen with the naked eye in *Futurama. Bolognium* features in both *Futurama* and *The Simpsons.*

12 Antimony and Timon, from the play *Timon of Athens.*

13 Sir Edward Elgar (1857–1934). The invention (the Elgar Sulphuretted Hydrogen Apparatus) became standard equipment in local schools around Herefordshire and Worcestershire. In his fifties, Elgar developed a passion for microscopes and his patented apparatus and microscope may be found at Elgar's Birthplace Museum, The Firs, in Worcester (see https://www.nationaltrust.org.uk/the-firs). At time of publication, ELGAR was also an acronym for a science project: The European Project for Gravitation and Atom-interferometric Research.

14 1F, 2A, 3D, 4B, 5H, 6I, 7C, 8G, 9E

15 Dilithium Li_2. In Star Trek, the fictional dilithium was given the atomic number 119. The real dilithium of course has the atomic number 3.

[§¶] 'transfusing emptiness into vacancy' was an idiom basically meaning that Borodin was wasting his time.

16 1E, 2B, 3D, 4A, 5C

17 Gaia, after the goddess Gaia who personified the Earth in Greek mythology.

18 Nobelium, No

19 1A, 2E, 3F, 4B, 5G, 6H, 7D, 8C[§‖].

20 Michael Jackson

21 Iodine, I

22 1D, 2E, 3B, 4F, 5A, 6C.

23 B, Heavy Water, D_2O

24 1E, 2C, 3B, 4A, 5E

25 Arsenic, As. For further information on the arsenic in wallpaper story see: *The toxicity of trimethylarsine: An urban myth*, Journal of Environmental Monitoring February 2005, William R Cullen and Ronald Bentley.

26 C Damien Hirst (b 1965)§**.

27 C Formaldehyde, IUPAC name methanal, CH_2O

[§‖] (All these poems may be found in *A Quark for Mister Mark*, 101 Poems about Science, Faber and Faber 2000).

[§**] The artist's own White Periodic Table from the Pharmacy reportedly fetched £112 500 at auction at Christie's.

28 B Copper sulfate $CuSO_4 \cdot 5H_2O$

29 C Henry Moore (1898–1986). The sculpture's original name was *Atom Piece.*

30 C Simon Mayo. To date, the books in the series are *Itch*, *Itch Rocks*, and *Itch Craft.*

31 B T S Eliot (1888–1965). Eliot was awarded the Nobel Prize in Literature in 1948.

32 B Wassily Kandinsky (1866–1944)

Periodic Table Puzzle

Wallace Carothers and the polymer Nylon

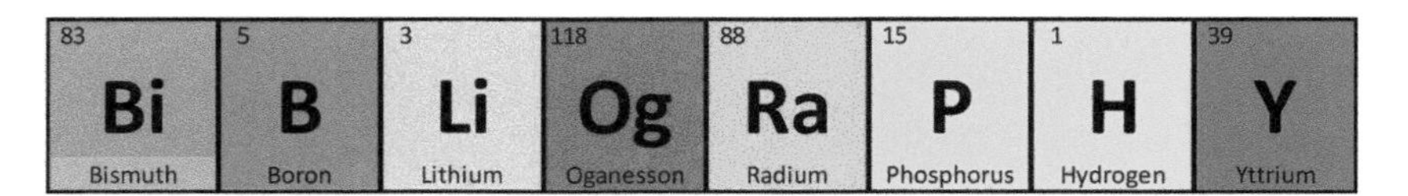

And further reading

As may be seen from the preceding contents of this book, it borrows from sources far and wide, but below is a selection of publications should any of the questions and puzzles have whetted your appetite to dig deeper (albeit one or two out of print).

- Suzie Sheehy, *The Matter of Everything: Twelve Experiments that Changed our World*, Bloomsbury Publishing, London, 2022
- Bonnie Garmus, *Lessons in Chemistry*, Penguin, London, 2022
- Simon Clark, *Firmament. The Hidden Science of Weather, Climate Change, and the Air That Surrounds Us*, Hodder & Stoughton, London, 2022
- Kathryn Harkup, *The Secret Lives of the Elements*, Greenfinch, London, 2021
- Cecily Wong and Dylan Thuras, *Gastro Obscura. A Food Adventurer's Guide*, Workman Publishing Inc, New York, 2021
- James M Russell, *Elementary. The Periodic Table Explained*, Michael O' Mara Books, London, 2019
- *Being Modern: The Cultural Impact of Science in the Early Twentieth Century*, ed. Robert Bud *et al.*, UCL Press, London, 2018
- Thomas Morris, *The Mystery of the Exploding Teeth and other Curiosities from the History of Medicine*, Penguin, London, 2018
- Stuart Farrimond, *The Science of Spice*. Dorling Kindersley, London, 2018
- Stuart Farrimond, *The Science of Cooking*, Dorling Kindersley, London, 2017
- *30-Second Chemistry*, ed. Nivaldo Tro, Ivy Press, London, 2017
- Paul Board, *Chemistry Crosswords*. The Royal Society of Chemistry, Cambridge, 2017
- *The Periodic Table Book. Our World in Pictures*, Dorling Kindersley India, Bharti Bedi, Senior Editor, This book is a good introduction for younger readers, 2017
- John Emsley, *More Molecules of Murder*, Royal Society of Chemistry, London, 2017
- Kathryn Harkup, *A is for Arsenic: The Poisons of Agatha Christie*, Bloomsbury Sigma, London, 2016
- Andy Brunning, *Why Does Asparagus Make Your Pee Smell? Fascinating Food Trivia Explained with Science*, ULYSSES PRESS, Berkeley, California, 2016
- *McCance and Widdowson's The Composition of Foods: Seventh Summary Edition*, Royal Society of Chemistry, Cambridge, 2014
- Richard Rhodes, *The Making of the Atomic Bomb, 25th Anniversary Edition*, Simon & Schuster, New York, 2012
- Jim Al-Khalili, *Quantum. A Guide for the Perplexed*, Weidenfeld & Nicolson, London, 2012
- Hugh Aldersey-Williams, *PERIODIC TALES. The Curious Lives of the Elements*, Penguin, London, 2011

- Sam Kean, *The Disappearing Spoon and other true tales from the Periodic Table*, Transworld Publishers, London, 2011
- Walter Gratzer, *Eurekas and Euphorias. The Oxford Book of Scientific Anecdotes*, Oxford University Press, Oxford, 2002
- John Emsley, *The Elements of Murder. A History of Poison*, Oxford University Press, Oxford, 2005
- Ken Silverstein, *The Radioactive Boy Scout. The True Story of a Boy who built a Nuclear Reactor in his Shed*, Fourth Estate, 2004
- John Emsley, *Nature's Building Blocks. An A-Z Guide to the Elements*, Oxford University Press, Oxford, 2001
- *The Merck Index. An Encyclopedia of Chemicals, Drugs, and Biologicals*, Maryadele J O'Neil, Senior Editor, Merck & Co Inc., New Jersey, 13th edn, 2001.
- Simon Garfield, *Mauve. How One Man Invented a Colour and Changed the World*, Faber & Faber Ltd, London, 2000
- *A Quark for Mister Mark. 1001 Poems about Science*, ed. Maurice Riordan and Jon Turney, Faber and Faber Ltd, London, 2000
- John Mann, *Murder, Magic & Medicine*, Oxford University Press, Oxford, 2000
- Cherry Lewis, *The Dating Game. One Man's Search for the Age of the Earth*, Cambridge University Press, Cambridge, 2000
- David, Ian, John & Margaret Millar, *The Cambridge Dictionary of Scientists*, Cambridge University Press, Cambridge, 1996
- *The Faber Book of Science*, ed. John Carey, Faber and Faber Ltd, London, 1996
- Primo Levi, *The Periodic Table*, Everyman's Library, Random House (UK) Ltd, London, 1995
- William H. Brock, *The Fontana History of Chemistry*, Fontana Press, London, 1992
- Harold Egan, Ronald S. Kirk and Ronald Sawyer, *Pearson's Composition and Analysis of Foods*, Longman Scientific & Technical, London, 1987
- W. E. Flood, *The Origins of Chemical Names*, Oldbourne Book Co Ltd, London, 1963
- W. E. Flood, *Scientific Words. Their Structure and Meaning*, Oldbourne Book Co Ltd, London, 1960

Web resources:

- Atlas Obscura, atlasobscura.com
- Bureau International des Poids et Measures, Home of the International System of Units (SI), bipm.org
- European Organisation for Nuclear Research, www.home.cern
- Explorations of everyday chemical compounds, www.compoundchem.com/
- Geological Society, www.geolsoc.org.uk
- Giant's Causeway, www.nationaltrust.org.uk/giants-causeway
- Kids' Chemical Solutions. Capturing Early Eager Learners with Chemistry Comics, https://kidschemicalsolutions.com/
- Institute of Physics, www.iop.org
- Interactive Periodic Table: www.rsc.org/periodic-table/
- International Union of Pure and Applied Chemistry, iupac.com
- National Graphene Institute, www.graphene.manchester.ac.uk/ngi/
- Nobel Prizes Official Website, https://www.nobelprize.org/
- ORAU Museum of Radiation and Radioactivity, www.orau.org/health-physics-museum/
- Royal Institution, www.rigb.org

- Royal Society of Chemistry, www.rsc.org
- The Royal Society, www.royalsociety.org
- UK approved additives and E-numbers, www.food.gov.uk/business-guidance/approved-additives-and-e-numbers
- UK Meteorological Office, https://www.metoffice.gov.uk/
- Wikipedia, en.wikipedia.org/wiki/Main_Page